LOVE
RULES

LOVE RULES

HOW TO FIND A REAL RELATIONSHIP
IN A DIGITAL WORLD

JOANNA
COLES

HARPER

An Imprint of HarperCollins*Publishers*

HarperCollins books may be purchased for educational, business, or sales promotional use. For information, please email the Special Markets Department at SPsales@harpercollins.com.

FIRST EDITION

Designed by Fritz Metsch

Library of Congress Cataloging-in-Publication Data has been applied for.

ISBN: 978-0-06-265258-4

18 19 20 21 22   LSC   10 9 8 7 6 5 4 3 2 1

Contents

Foreword

If you've picked up this book, it's probably because you're frustrated with your love life. You may already have a partner, but you suspect he or she's not the one. Yet you don't quite have the nerve to break it off. What if it's simply the best you can do?

Or you may be single, divorced, or widowed and longing for love, trying everything possible to find a partner, turning over every digital stone—Tinder, Happn, Hinge, and Bumble—to uncover matches. You may have spent several hours filling out a thoughtful profile on eHarmony or a snappy one on Match .com. Or asked all your friends to set you up on blind dates. You have pored over the Facebook profiles of all the people you had crushes on in your past—and have even instant messaged a few.

But still you keep striking out.

You may find yourself in a protracted cycle of excitement and disappointment—meeting someone, going home with him or her, and then waking up with dater's remorse and wondering what on earth you're doing there anyway as he nudges you to say he's sorry but he needs you to kinda leave now because he has an early start. Or you may be feverishly fast-forwarding, imagining the destination wedding by the end of your first drink and the names of your children by the second. And then you never hear from him again. No text. Not even a three-second Snap. Total ghost.

Meanwhile, life for everyone else seems to flow merrily by, as your friends get engaged, get married, have babies—not always in that order, of course, but who cares. They have something going on. And you feel left behind.

Or you may have once been that person you now so envy—the one with the long-term boyfriend. But you broke up with him after college to play the field, and now, at twenty-nine, you're wondering if you made a mistake. He's engaged; you're still single, and no one has come close to treating you as well as he did. Or perhaps you were that comfortably married one who paired off early, but it didn't work out. And now here you are again, whether age thirty-two or fifty-two, having to get back out there. This was never part of your plan. Even more unimaginable, you didn't realize that when you eventually posted your online-dating profile and leaned in to see the matches, you would find a Niagara Falls of dick pics, or even worse, their video versions. Online dating is not for the faint of heart.

The landscape has changed so radically in such a short period of time. Not just how we meet potential partners, but how honest women are finally being about how harrowing that can also be. Leaders, icons, and figures we once trusted—across all walks of life—have been brought down with startling velocity by sexual harassment charges, and women, who for years felt crushed by an impossible silence, have found their voices to tell sometimes terrible stories of bullying, harassment, and humiliation. This is another reason I wanted to write this book—to help women navigate what can feel like very tricky, even dangerous, terrain.

We all know that finding love is possible: the story you

heard about the friend of a friend who married the Nobel Prize winner she met on JDate . . . It can and does work out, right?

Hang in there, whoever you are.

This book is for you.

This is a diet book for love.

Food and love have so much in common. We have huge appetites for both. We can't live without them. But not all food is created equal, and neither is all love.

Just as there is junk food, there is junk love. And like junk food, junk love is fast, convenient, often attractively packaged, widely available, and superficially tasty. But the calories are largely devoid of real nutrients and leave you hungering for more, even as you smart with a lingering sense of shame after a binge. And yet, both junk food and junk love require enormous amounts of willpower to resist.

Amid the clamor of conflicting dietary advice, one book stands out to me for its clarity. Michael Pollan's *Food Rules* is a straightforward and gimmick-free guide to eating well.

In *Love Rules*, I hope to do for relationships what Pollan did for food. To help women navigate their love lives in this very modern, fast-paced, and—what can feel to many—incredibly lonely digital age. In this case, social media and online-dating sites are equivalent to the Walmart and Costco of our relationship lives: You have to wade through rows of Little Debbie snack cakes and family-pack bags of Cheetos in order to find the aisle of fresh produce. The organic apples and almonds are there—you just have to know why they are important and then where and how to find them.

Substitute apples for a real-life healthy relationship, one

grounded in intimacy and trust—with someone who makes you feel good about yourself. Someone who will stand beside you for the long haul.

To that end, I've created a series of guidelines. These are rules informed by my many years working at women's magazines, including *Cosmopolitan* and *Marie Claire*. They have been shaped by extensive consultation with psychologists, social scientists, anthropologists, doctors, college professors, therapists, and religious leaders, as well as by conversations with the many young women I have worked with in the media world. And, of course, those I have met over the last decade as a result of my work. The thousands of smart, independent, successful women who had so much going right—work, bank, friends, wardrobe, 401(k)s—but otherwise felt unhappy or unlucky in love. They have created enviable lives for themselves, yet they are missing someone to share that life with and maybe have a child or several with. That is the one thing that eludes them. Or, they may have two kids and a failed marriage, and are wracked with worry that they may never meet anyone to share their life with again.

The rules that emerged from these conversations are essentially love hacks.

This book will enable you to identify what you want in a relationship and when you want it. In order to do that, you have to turn inward. This book is truly about finding yourself and cultivating your self-worth in order to find the right person to share yourself with. And to do that well, you have to apply to your love life the same often fanatical focus that you devote to losing weight, getting fit, or changing your job.

I have used mainly he/she and his/her throughout this book, but however you identify and whatever your sexual orientation,

Love Rules applies to everyone. So please, wherever relevant, just substitute your own preferred pronoun.

And on a somewhat related note: you will see that some names have been changed in the case studies that follow. As an editor, I don't generally like using pseudonyms—it has always made me feel that people could exaggerate whatever they were saying—but the very personal nature of some of the stories and case studies in this book meant a pseudonym was necessary to protect privacy. We fact-checked each story but agreed to give certain people the right to change their names.

IT'S TIME TO GET REAL

Approach this as you would begin a diet. That means counting every calorie both in and out. A calorie is a unit of energy—so how many times do you check your Tinder account? Or how many hours have you spent on Facebook bingeing on ex-partners' posts? These are emotional calories, and they use up your positive emotional energy that would be better expended elsewhere.

You need to take an emotional calorie count.

How many times have you struck up a conversation with the good-looking guy at your local coffee shop? When was the last time you had great sex that left you feeling satisfied and connected? Or a relationship with someone who made you feel truly good about yourself? These are the calories that will sustain you.

You need to take an emotional calorie count.

And then what about the calories you don't even realize you are consuming? The equivalent of those french fries snuck off someone else's plate at dinner. You didn't order the fries yourself, but you ate them just the same. These are the incidental calories that satisfy in the moment but can do longer-term damage that's surprisingly hard to shake off. The random hookups that may seem fun or daring at the time but leave you feeling annoyed with yourself. The online sinkholes we all have fallen down—in sweats on a Saturday night, hearting the Instagram shots of friends out partying and clearly having more fun than you, or comparing your sad inner life with the obvs perfect external one that your colleague at work curates on Facebook. These binges may feel beyond your control in the moment, but they likely are not leading you to find love.

Psychologists look for pattern recognition as the key to understanding and changing behavior. With the help of *Love Rules*, you can discover your intolerances and allergies and be your own relationship nutritionist.

In the same way that so many of us are used to keeping food journals, recording every morsel we consume in order to understand our food habits, it's time to keep a love journal. A consistent log of your own behavior can highlight your repeated patterns that lead nowhere—and help you identify the triggers that cause them.

We all know that there is almost nothing more draining than a bad relationship. And there is nothing more life enhancing than a good one. It makes everything else look, feel, and taste better.

PART ONE

TAKE INVENTORY.

Rule #1

Establish your ideal love weight.

It's time to strip naked and look in the mirror. Ask yourself, "What do I want in a relationship?" What is your ideal scenario? Be honest.

Who is your dream catch? And what may be more realistic?

We all have an ideal weight. When we hit it, we feel happy. Sexy. Confident. And when we are five, ten—or forty—pounds away from it, we feel disproportionately terrible. It can feel as though we just can't get back in control, and it often makes us crave unhealthy food and want to eat more. So first, take a hard look at yourself and your current love life.

It's similar to getting on the scale. How much do you weigh? Is that your ideal weight? If not, what is? And again, be realistic. So if you are five feet six and 152 pounds but want to be 120 pounds, ask yourself, "Have I ever weighed that?" What is the lowest you have been? What did you have to do to maintain it? Maybe 135 pounds is more realistic—and healthier—for you.

Now swap that for relationships. Sure, your best friend is dating a guy you think is "perfect"—but for whom? Hopefully for her. What about you? Who is the 135-pound equivalent of your potential partner? Not the John Legend or Ryan Gosling 120-pound version, but the realistic one. Perhaps even the guy who works in the IT department at your office and leaves flirtatious sticky notes on your desk or signs his work emails with a winky emoji. Or the shy philosophy major in your dorm who clearly has a crush on you. You agree with your friends that he is a dork but have found yourself wondering about him all the same. And actually you feel good when he's around.

Forget about what anyone else wants for you—your mom, your best friend, your sister, your colleagues, your aunt, or your neighbor—think about what *you* want. This is oddly difficult to do. We are constantly seeing ourselves through others' eyes; it's human nature. The phrase "looking-glass self" was first coined by the sociologist Charles Horton Cooley in 1902 and describes this phenomenon, in which we actually define ourselves by our interactions with others. That mirror is magnified a thousand times in our modern world as there are so many points of comparison between ourselves and others. It can lead us off track. As the pioneering cyberpsychologist Mary Aiken, author of *The Cyber Effect*, puts it, "We spend all of our time investing in trying to understand our 'self' from the feedback from others rather than actually knowing who we truly are."

So the first rule is to start thinking about who you truly are—and what turns you on or off, thrills you in the moment, and lasts for the long haul, because that is the key to finding a sustaining relationship.

And so again, ask yourself, "What do I want in a relationship?"

There is an ancient Greek expression that we need to make modern again: Know thyself. So much of life today is spent comparing ourselves to others—whether that "other" is your best friend, the lawyer who just married a tech entrepreneur and is already pregnant with her first kid at twenty-nine. Or your colleague who got a raise instead of you and is on her third date with the hot guy she met on Tinder. Or any one of the improbably nice Kardashians. Everyone is so busy looking at, liking, and idealizing other people's lives that we each define ourselves and our desires and goals in reaction to them, versus the internal deep work of asking, "What makes *me* truly happy?"

In a restaurant, we may ask for suggestions, but we don't let others tell us what to eat; we choose from the menu ourselves.

Everyone has different wants and needs, most based on past experiences and future aspirations. You might really like the shy, quiet guy who works in accounting—the one who wears a zipper cardigan and, gasp, Merrells. But your best friend thinks you should date the chatty trainer who flirts with you at the gym. Going along with what other people think is best for you—but what does not feel right in your heart and gut—is not what we are going for here. In a restaurant, we may ask for suggestions, but we don't let others tell us what to eat; we choose from the menu ourselves. Bat away the white noise and the cultural pressure. Ask yourself, "What do I want in a partner?" You need to choose for yourself first and worry about the peanut gallery later.

TO DO

Establish your ideal love goal.

(Fill in the blanks.)

I want to find

_____

My ideal partner has the following three qualities:

_____

_____

_____

Analyze what others say they want for you and check it against what you want for yourself.

Parents:

_____

Best Friend:

_____

Siblings:

_____

Colleagues:

_____

Online Friends:

_____

Do their expectations for what you deserve in a partner align with what you want?

_____

If so, how are they the same?

_____

If not, how are they different?

_____

Rule #2

Clear out your cupboards and sweep the fridge.

Once you decide what you want in a relationship, you must make an active plan to achieve it. As with any successful diet, that plan starts by setting realistic goals and continues by sticking to them and monitoring them. And if it's not working—if you find yourself cheating or slipping up—then be brutally honest about why.

It's time to start tracking the data on your own love life and then review the results. Be your own data analytics expert.

We live in a culture where we can now track everything—our daily steps, our REM sleep, our carb intake, our pulse rate—so it's time to start tracking the data on your own love life and then review the results. Be your own data analytics expert.

Take this quest as seriously as you take any other item on your to-do list, whether that item is finding a job, losing twelve pounds, training for a 10K, or paying off your credit card bill.

So, where to start?

Buy a notebook to dedicate to this one thing. Sure, it sounds old-fashioned, but studies prove that you retain more information by physically writing it down, pen on paper, than tapping it out on a keyboard. Of course, if you really can't imagine writing by hand, then you can always start a diary on your computer or iPad. Either way, make this your purpose-driven love journal and make writing in it a ritual that takes you outside your day-to-day to-do lists and other "notes to self."

The point is to do this in a way that takes your quest seriously. For this reason, I much prefer a quality notebook, not a yellow lined legal pad with disposable pages, but a beautiful notebook, one you will enjoy opening every day. I have dozens of notebooks, big ones, tiny ones, some I've written in until they're full, others completely empty, still waiting for their purpose to hold ideas and promise. I give them as gifts, and I am never without one. You never know when you will need to write something down—a quote, a thought, an idea, a dream, an ambition. But this particular notebook has a purpose. Title it however you like: maybe with a favorite quote or line from a song, or a saying from your favorite icon. And then find a private space to keep it, where no one else would think to look, so you alone can read it.

Give yourself an hour, maybe even pour yourself a glass of wine or go sit in a café, and then on the first page, answer the following question:

What do you want more of in your love life?

The prompts below are just suggestions; you will have a wide and varied list of your own.

Fun sex with no obligation to call them afterward?

More laughter? More trust?

Someone to travel with? To explore with?

Someone to Netflix and chill with?

Someone you know you're going to hang with so you don't have to think about actually planning what you're doing on Saturday night?

Someone to help you get over the heartbreak of your most recent breakup?

Someone to marry in the next year or so and start a family with?

Someone who will come to those work social events you find so awkward and be a better half?

Give yourself the time to really think about your answers. The more you know about yourself and what you actually want, the closer you are to finding it.

Having thought through, and written down, what you want in a partner, at least gives you a guide. Love is unpredictable and surprising. It can catch us off guard. We can't predict precisely when we might meet the love of our life or how many hoops we may jump through along the way. It's both exciting—and terrifying.

It feels out of control not knowing when a relationship might appear. And it's so frustrating. If only we knew that in six months or even two years we would meet someone, then we could relax and enjoy the lead-up to it.

My guess is that you have a job, a place to live, a closet with enough clothes, that you care about your physical appearance and fitness, and that you have a network of family and friends. We launch our adult lives after college and work so hard to achieve all these things. So why does finding a partner, a person to share all this with, feel so daunting?

The twenties are supposed to be among the best stages of a woman's life. You're done with college, ready to explode into the world of work and continue exploring your sexual self, only with a slight edge now as you embark for the first time on a life where you alone—no longer your parents or your college—get to create the boundaries. It's a time of self-discovery and adventure, and if you're working, then you have a bit more money to try new things, too.

Plus, you're at your sexual prime, widely reported to range from your midtwenties to thirties, and so of course you want to experiment. It's great to have fun. But if you ultimately want a sustaining relationship, as many women do, then it's okay to acknowledge that.

Giving yourself permission is half the battle. "It's so hard for women to admit that they want this," says Helen Fisher, PhD, the renowned biological anthropologist and author of many books on love, including *Anatomy of Love: A Natural History of Mating, Marriage, and Why We Stray.* "Many find it too retro, but the truth is that finding your life and/or mating partner is hardwired into all of us. It is the ultimate prize."

Inevitably, what can feel like a prize in your twenties doesn't always feel that way in your thirties or older. For those whose marriages have not worked out, the new dating world of potential partners can feel like a hostile environment. My point is, it doesn't matter where you are in your life or how old; figuring out what you want and need in a partner at this very specific time in your life will help focus your search.

In the same way that opening your bank statement can actually feel like a huge relief, stopping for a moment to consider your relationship goals will make you feel that much closer to achieving them. Who hasn't gone hopefully home with someone and then been crushed by the announcement, usually right after sex: "That was great, but just to be clear, I'm not looking for anything serious, and I hope you understand." You nod and pretend to agree. You may even say, "So relieved you said it first! I don't want anything serious, either! I've got so much going on!" And then you sweep up your dashed hopes, struggle back into your skinny jeans, and pretend you don't care. But you do.

It's time to be honest with yourself and with others.

Consider how comfortable we feel doing this with other areas of our lives. We place our ambitions and desires for that perfect job or career under a microscope. We spend four years in high school worrying about what to study in college and then another four years in college worrying about if what we're studying will get us the career we want. If someone doubts us, then we accuse them of not being sufficiently ambitious for us. But we don't examine our love lives with the same scrutiny, other than with our friends over too many glasses of chardonnay, possibly the least objective teachers ever! We have no classes or schooling on relationship intelligence because we are somehow supposed to have that under control, innately, as if circumstances will one day organically present us with the perfect partner. And yet, it is much easier to change jobs or offices or full-on careers than it is to change romantic partners, especially once you have kids together.

My friend Denise* was thirty-five, single, and wanted a family. She had just sold her travel company to Travelocity and

---

*Denotes pseudonym.

realized that she could not "keep hitting the snooze bar" on her biological clock. So she sought out a partner with "intention." "It was unromantic," she says.

She went to a Jewish singles happy hour with a friend and no expectations. "We were both single and decided, if it's boring, then we will go out for dinner," she says. That was in 2000, and where she met her future husband. He suggested hiking on their first date. She wore a T-shirt, shorts, and sneakers, and thought, "I want to meet someone who is going to accept me for me," she says. "And I remember thinking, even if this does not work out, I'll get a nice hike in."

It did work out—and Denise wasted no time. The two married in 2002 and had Katie in 2004. "I'm a businessperson— when I put my mind to something I can usually achieve it," she explains. "That was my insight: I had to treat finding a partner and having a family like my approach to my career."

In my experience, Denise is the exception, not the rule. While our relationships need as much scrutiny as our working lives, they often don't receive that attention until it's too late. It's so easy to get caught up in the romance and relief of a new relationship that we sometimes miss the signs that we should hold out for someone else with whom we have more in common—for someone who makes us truly happy when we're with them, or at the very least, respects us.

Our relationships need as much scrutiny as our working lives, yet they often don't receive that attention until it's too late.

It's frustrating that this is so hard to do. Women are more "equal" than ever before. We're "leaning in" in the boardroom, demolishing the gold medal record at the Olympics, and out-graduating men in higher education. (College undergraduate intake is now 57 percent women, 43 percent men.)

I have spoken to thousands of young women through my work, and what has surprised both me and them is that, despite their smarts and ambition and capabilities, they are more confused than ever about love and relationships. Sex is widely available, sex without judgment, too. Hooking up is easy, though many have told me that it's not as fulfilling as they expected. The refrain I hear over and over is "I love my life. I love my job and my friends. I just never expected to find myself single. I just never expected to find myself in this position."

So if you do want a relationship where you can talk confidently about having a future as part of a couple, now is the time to admit it. Fess up. You are not alone. And once you have set your bigger goal—whether to be married with children by your midthirties or to be in a serious relationship by the end of this year—you can start to focus on the tactics that are necessary to reach it.

TO DO

Establish a regular habit of writing in your love journal, and answer the following:

Have your love goals changed since you did the to-do items for Rule #1?

If so, what are your new goals?

Have you made steps in achieving them?

Identify your triggers: What do you think has gotten in the way of finding love?

Start this quest by being honest and conscious of what you want, and then ask yourself what gets in the way of it.

We have all had those moments when we saw the cookie, promised ourselves we would just eat half, and then went on to eat four and felt ill. For some, this becomes a daily habit. In order to find the best dating diet for you, first you must know your weaknesses.

Rule #3

Begin a dating detox to reset your metabolism.

Whether you need to lose weight or simply want to get healthy, it's always good to start with a thorough intake. Most women I know have a giant invisible calorie counter hanging above their heads at every meal, and they apply their own set of accounting rules. It's food math, and don't tell me you don't do it. You wave the bread away and hold the fries so you can have two mojitos instead and feel virtuous. Sound familiar? You skip carbs but then double down on protein and chocolate and find yourself still hungry, with a headache. Out of balance.

We lie to ourselves about our emotional intake, too. You reason that sleeping with an ex doesn't really count because it's not as though you're increasing the all-important *number* of men/women you've slept with. What you don't calculate is that it still adds up to emotional calories—your inevitable inner conversation about what his new girlfriend would say or about why you split up in the first place. It's not moving you forward; it's taking you backward. And it's all energy that could be better deployed elsewhere.

A good nutritionist will want to know your daily food habits. Do you applaud yourself for your self-restraint around all food until 4:00 p.m. and then become unhinged, scarfing whatever you can find until you feel so depressed you renounce food forever—until you start the cycle up again the next day? Do you restrict yourself to one chaste glass of wine a night, which you sip obsessively on the quarter hour? Or do you convince yourself that splitting a bottle is so much more economical than buying by the glass?

Nutritionists will also grill you to find out if you actually consume enough healthy calories to keep your body satisfied and functioning properly. And they will certainly ask about your medical history—such as, does obesity run in your family? Have you ever had an eating disorder? What is your BMI?

Do the same thing for your love life.

In your real (or digital) notebook, answer the following:

Have you ever been in love?

Who was your first love? Was it mutual?

Did he (or she) treat you well?

Did you treat him (or her) well?

How many people have you dated? For how long?

How did each relationship end?

What types do you go for?

How many long-term (more than three months) relationships have you been in?

Do you like monogamous relationships?

Have you ever cheated (be honest)? Why did you cheat? How did it make you feel? How did it end?

What is the happiest you have ever been in a relationship? What about it made you happy?

When is the last time you felt that way?

The next step is to take an inventory of all the partners you have had in your past. Include the ones that didn't work out as well as the ones you had successful relationships with. Open all the cupboards to your past. Clean out the deep freezer. Do you stockpile Diet Cokes but then sneak Ben & Jerry's ice cream bars on your way home from work? Do you say you are going to exercise and then decide that you feel a cold coming on and might be better off taking it easy? Some of us prefer salty, gorging on bread and cheese; others go for cake pops. Others pride themselves on eating only cottage cheese and carrots for lunch but then devour donuts as 4:00 p.m. comes around.

All these patterns apply to relationships as well.

Just swap relationships for food.

Whether you like the bad boys, or you have a soft spot for the supersensitive ones, or you ping-pong between girls and boys, or perhaps you've only had unrequited crushes, you likely have a relationship pattern that you need to examine in order to determine the right habits going forward. You need to admit your weaknesses and cravings—and single out your strengths, and moments of happiness, too.

Most diet books promise a thinner, fitter you in ten weeks.

This book doesn't promise that, but if you want a chance at finding a happy or fulfilling relationship, then you have to first recognize your patterns. Include every one-night stand, every online dalliance. Don't ignore the unrequited crushes, either.

If you want a chance at finding a happy or fulfilling relationship, then you have to first recognize your patterns.

Write it all down—every person you have hooked up with, spent hours obsessing over, or cheated on. Just as you have to get on a scale every morning or look at yourself in the mirror at the start of every diet, you must begin your dating detox by being honest about your past relationships.

And that includes your role in each relationship. It is time to ask yourself why you are single.

An understanding of how you present to the rest of the world is key. Before you begin the dating diet, imagine what it would be like to go on a date with yourself. How do you come across to others? If someone Googles you beforehand, what will they find? What image do you present to the world? Consider the following:

> **Do you only post photos of yourself when drunk? Or with other drunk friends? Or doing exciting things that other people might not have access to?**

> **Do you keep up with the news?**

Do you find yourself to be the best topic of conversation? In other words, do you only talk about yourself?

Are you compassionate?

Are you a good listener?

Do you come across as bone-crushingly ambitious?

Take a selfie and for a moment get outside yourself.

What would it be like to date you?

And if this is too hard, then ask your parents or sister or a friend. Treat this like a 360 review at work. Ask people you trust to kindly tell you what your strengths are and what you could work on doing better. You don't have to agree with them, but they may give you insight you can build on.

Next, compile a list of the blowouts that you've had with any of your exes. Does a pattern emerge? Is there a narrative thread? A common complaint about you in relationships? Too selfish? Too messy? Irreverent? Too thrifty? Too harsh? Not enough fun? Too much fun?

Is there a consistent criticism that keeps coming up? Something that you are supersensitive to? If so, perhaps it indicates an element of truth.

This reminds me of my friend Danielle*, who did not realize that she was lactose intolerant until her early thirties, when she got terrible food poisoning that landed her in the ER and then in the hospital for several days. The culprit was a pork chop at a Chinese restaurant, but as she slowly started to eat again after three days with no solids, she discovered that she was having a strong reaction to milk and dairy products. The reaction to the bad pork chop revealed her growing intolerance to dairy. "I had forgotten what 'good' felt like because I was so used to being uncomfortable and bloated," she says. "It was only after I had stopped eating entirely and then started reintroducing foods that I could tell immediately what threw me off."

Food intolerances are a good parallel for dating.

Following the notes you have already made, go back and add all the unrequited loves, including your very first crush in kindergarten. Mine was Andrew, whom I met in primary school. He had a blond pompadour and had somehow appointed himself class leader, and we basked in his attention. I can still remember the thrill of being rose queen to his king at the annual summer pageant. He was confident and funny, and until writing this book, I hadn't thought about him since— and I am sure he hasn't thought of me! But looking back, he unwittingly set the template for the guys I found attractive, and I went out with several like him. It's important to remember every person you have fallen for because you will likely see a pattern of the types of people you find attractive—quiet, noisy, funny, retiring; rabble-rouser, outsider, rule-follower, leader—emerge.

Write down your first crush's name. Was that a he or a she? How old were you? How did the attraction begin? How did it end? Then answer the following questions:

How would you sum up your dating history?

Are you a serial monogamist?

Someone who can't commit?

Do you think of yourself as lucky/unlucky in love?

How many people have you had sex with? How many of those from online encounters? How many sexual encounters with total strangers?

What is the best part of having sex with someone you don't know?

What is the most enjoyable part of having sex with someone?

If you are currently in a relationship and unhappy, why do you stay?

What are you scared of?

How much do you change your personality around your boyfriend or girlfriend?

Have you ever had an experience where you were not sure if you had been assaulted? If you have been assaulted, write that down, too. It is far too common and one of the reasons I wanted to write this book.

Write down every encounter you can remember. Include each obsessive Facebook crush—note how many minutes you spent going through his vacation photos or counting how many times that other girl liked his Instagram photos, too. While you're at it, add up all the time you spend waiting for your boyfriend to want to (a) move in, (b) propose, (c) move out of his mother's house into his own place, and (d) take you somewhere you actually long to go.

Stop complaining about how dating sucks and start analyzing what is actually going on in your love life that makes it feel like it sucks.

Look at the data.

Just as it's easy to get into bad food habits, it's easy to get into bad love habits.

Take Sally, the daughter of a friend and a sophomore at a smart liberal arts college. One evening over dinner, I asked her what her weekends were like. I wasn't expecting her answer.

"Well," she replied casually, "my friends and I all go out on Friday nights, get drunk, and hook up. And on Saturday morning, we go down to the health center together to get Plan B."

I knew the over-the-counter drug was easily available— sold in vending machines on college campuses and handed out at wellness centers—and that many young women are not taking birth control pills because they worry that the regular intake of hormones is bad for their health. But I was still gobsmacked by Sally's blasé admission, and worried for her. Here was a highly educated and intelligent girl I had known since she was skipping off to primary school in the mornings. Now she was admitting that she and her friends go out together knowing that any one of them may have sex with a stranger and without protection, leaving them open to both STDs and pregnancy. "Plan B" almost sounded like a badass badge of honor, as in, "Oh God, I was so drunk last night, I don't even know if I had sex!"

I was also unsettled by how grim those Friday nights

sounded. Pregaming and then having passed-out sex—was this the new fun on the campus? #SquadgoalsPlanB&brunch! Did no one in her crew say, "Actually, this isn't fun, and it's probably not doing us much good"? I stuffed it in my mental drawer to unpack with the *Cosmo* staff later, and then over the next few weeks, I started asking other young women if that scenario sounded familiar.

It did.

It sounded very familiar to the interns at the office, to readers, to women who wanted to tell their stories about what many relationships are really like for women in their twenties. And to Leah Fessler, who based her senior thesis at Middlebury College on hookup culture and then wrote a remarkable essay on her findings for the online journal *Quartz* titled "A Lot of Women Don't Enjoy Hookup Culture—So Why Do We Force Ourselves to Participate?" It went viral.

That was May 2016, and the line that stood out was "The fact that most of these guys wouldn't even make eye contact with me after we had sex or would run away from me at a party was the most hurtful thing ever."

As we commissioned more and more stories at *Cosmo*, I found myself asking: Where are the good sex stories? Where are the stories of glorious obsession? The stories of blissful romance, of longing, of waiting for the person you can't stop thinking about to notice you or text you or—dare you even think of it— agree to meet?

And most important, why wasn't there any mention of the word that haunted my own teens and twenties—love?

There is so much talk of connecting these days yet such little trust in those connections. There is so much ease to hooking up yet not always much pleasure in it. The goals women are under pressure to accept have changed from love to sex. We live in a world where men and women can say, "I want to have sex tonight. Let me go online and figure out who's out there." There have never been more options available in finding a partner, whatever your age or stage. This is a new moment.

There is so much talk of connecting these days yet such little trust in those connections. There is so much ease to hooking up yet not always much pleasure in it.

And in this new world, women need to be very clear with themselves about what they want out of the deal.

So in your relationship diary, record each Snap that leaves you feeling breathless, every Tinder swipe that has you hopeful, every coffee shop flirtation that gives you butterflies. They all matter.

You've already begun the process of writing things down, and my guess is that some early emotional relationship truths have emerged as a result.

Do you always fall for the guy who ignores you?

Do you stay with people who make you feel bad about yourself?

Do you find yourself suppressing what you really want for fear of frightening him or her away?

Do you expect more from your love interests and pretend that you don't mind when they don't deliver?

Once you have thought hard about your past, think to your future.

Who do you think is truly the best type of match for you?

What are the characteristics you feel you need in a partner?

Does he or she need to be good looking? Kind? Financially stable? Artistic?

Make a detailed list of the ten qualities that you are looking for and don't forget to include your thoughts on children. Do you want them, have them already? Are you someone who has never had a pang and is happy to be child-free? Whatever your stance is, state it in ink (or tap it out on a keyboard).

What are your requirements? Your deal breakers? And why?

Do you always go for the moody introverts who get jealous when you want to see your friends? Do you develop mad crushes at work but are too shy to even talk to any of them? Or do you keep hooking up with your ex because he lives in the same apartment building as you and it is easier than having to put yourself out there?

What are your patterns?

Identifying your unhealthy patterns is the first step to breaking free of them. Just like you might crave a brownie to go with your skim latte every day around 4:00 p.m. Or you cannot imagine seeing a movie without inhaling a large tub of popcorn doused in fake butter.

What are those cravings and weaknesses when it comes to love? And how can you break them?

And while you are at it, this is a good moment to take stock of your friends and do a friend cleanse:

Analyze *all* your relationships—not just the exes who left you feeling devastated or the compelling but infuriating FWB who refuses to take it further than sleeping with you on Tuesday and ignoring you the rest of the week, but also the friends who assure you, as you angst over whether you're doing the right thing, that "at least you're having sex."

Friends really do come and go throughout your life. And friendships do change as you develop. Cut out the friends who make you feel good in the moment but bad on your way home or the next day. Instead focus on those who make you feel good about yourself and give you good energy, the fruit and veggies of your life.

It's okay to shed people.

Someone who sustained you at twenty-two does not necessarily have the same appeal at thirty-five. This applies to both love interests and good friends. Our female friendships are hugely important, the stuff of life, but they do not necessarily help in the romantic space. They can also thrive on your drama.

Friends can get jealous or insist they know what is best for you. They can even sabotage potential partners. So do a friend check: Do you have friends who support your bad decisions? Who are enablers? Or commiserators? Who make you feel out of control? Do your friends collude with you to claim you are just too intimidating for most men, when in fact you can (on occasion) be loud and boorish?

Extend your detox to your friends.

So extend your detox to your friends. I'm not suggesting one of those extreme cleanses where you're supposed to survive on water infused with cayenne pepper for ten days and you end up delirious. But it's worth suffering the headaches that come from a decent detox as your system rebalances itself without the usual junk you have become reliant upon.

It's time to move you forward.

PART TWO

DATE. RINSE. REPEAT.

Rule #4

The treadmill won't run on its own.
Climb on and press Start.

Just like the Fitbit tracker you bought hoping it would somehow melt away your muffin top, dating apps are mere tools in your arsenal to find a mate. A BOSU ball can't do the crunches for you. A flirty text exchange with a Tinder match might give you a quick dopamine high, but it's not a comparable substitute for meeting in person or picking up the phone and hearing someone's voice.

Dating apps are genius because they will help you locate potential partners. But only you can figure out which ones are worth meeting. "These are not dating sites, these are *introducing* sites," Helen Fisher says. "The only good algorithm is your own brain."

Sean Rad, cofounder of Tinder, agrees. "Nothing replicates real life," he says. "There is no substitute."

Internet dating gives us the illusion that dating is easy. Its speed and access gives another false sense that you will know instantly, when you meet this person in real life, if he or she may be worth a second date. But the idea of love at first sight is deeply flawed. We all have friends who have suffered the indignity of dressing up to meet someone promising from an online match and then, in real life, having that person arrive at the table, give them a quick once-over, and say, "Hey, you look great, but I just know this isn't going to work and so let's not waste our time." The truth is you can't possibly tell if someone is going to be a match in three seconds.

"In your real life, you might meet someone at work and think they are a complete idiot for a year," Fisher says. "And then over time, you discover they are kind, funny, and love to play tennis and you do, too. This natural system of getting to know someone is being killed by anyone who expects to have instant romance on the first date." It feels counterintuitive in our hyperconnected and superfast world—uncomfortable, too. But it is during those awkward getting-to-know-one-another moments in real-life places that love starts to blossom.

Rad started Tinder to replicate a coffee shop or bar experience—minus the awkwardness. "The core thesis behind Tinder is that the fundamental thing that prevents people from walking over and saying hello to someone that catches their attention is that there is no context," he explains. "So there is not a socially acceptable reason to go say hello. The timing might not be right, the circumstances might not be right. Those are external. The internal mechanisms at work are humans' fear of rejection."

It is far easier to settle in on your sofa and swipe through profiles, either with friends or alone. You can scan hundreds of possibilities, write a profile to poke or message them all, and respond to those who do the same to you.

But you still have to meet that person and have a real connection if these tools are going to work, which is why you should try the gamut of dating apps and settle on the one that feels right to you, whether it's Tinder, Match, Hinge, Bumble, Coffee Meets Bagel, JDate, or any of the other dozens that are out there.

You still have to meet that person and have a real connection if these tools are going to work, which is why you should try the gamut of dating apps.

Apps are like cars; they need to be handled with care. They are unparalleled for broadening your horizons and taking you on journeys you might not have imagined, but you need to proceed with caution. You need to know what gear you're in and to make sure you signal exactly where you want to go so others know, too.

You also need to banish any expectations that an app will lead you directly to Prince Charming, as he does not exist (see Rule #15). As Esther Perel—the brilliant couples therapist, author of *The State of Affairs*, and host of the Audible podcast series on love, sex, and relationships *Where Should We Begin?*—points out, that's too much pressure to place on one person anyway. "The more freedom you have, the more relationships are like a free choice market, the more you also have to deal with the uncertainty of knowing that this is the right person," Perel says. "The more freedom you have, the more you are also riddled with self-doubt and uncertainty. You knew what kind of person you needed to bring home, what social class, what education, what religion, all those categories, which were pretty much selection processes, mating categories. Now the world is wide open and you have to reassure yourself."

EXPAND YOUR NETWORK

This is the dilemma of being so overly connected. It puts too much pressure on the hunt for this very specific person who will be the answer to your dreams. Thinking of dating apps in this way is unrealistic and will psych you out. Instead, use these apps to expand your real-life social network. Dating apps are best understood and used as introducing tools. They can introduce you to people who may, in turn, introduce you to someone—maybe their best friend, brother, colleague, or cousin—who you decide is special. These apps are a powerful way to hone what Rad calls "an innate humanistic desire to meet other people and expand your social network."

I love his use of "social network" here and have always felt dating apps should do more with this idea. Because their real power is in how they can create the possibility of meeting people already on the margins of your world whom you might otherwise miss connecting with because you aren't normally in the same place at the same time. Think of all the people you have met and your delight when discovering you were actually at the same party two years ago but didn't meet then, or you were holidaying on the same beach during the same spring break, yet you didn't run into each other.

So instead of thinking, "How do I find a partner?" think, "How can I expand my connections? How can I create more possibilities in my life, whether for relationships or friendships?"

Reimagine dating apps as a way of meeting more people as opposed to meeting "the One." The chances are you won't be able to spot the One with a quick match anyway. But you may meet someone who is worth meeting for a quick drink. And he may have a best friend or brother that you end up dating for six weeks, or you might introduce your friend to his brother.

Dating sites are in fact beginning to do this more. Tinder Social was launched in 2016 as a way to meet friends and connect with other people who want to spend the evening or afternoon either going to the beach or a concert or on a road trip. To use the app, you create a small group of your existing friends. Tinder Social then matches you with other groups and, just like regular Tinder, you swipe right or left to join other parties to hang with. So one group may invite the other(s) to go to a beer festival or to play volleyball in the park that Saturday.

This is where tech really works. With work life taking up so much of our time during the week, our ability to meet both potential partners and new friends is often limited to Friday or Saturday nights, depending on where you live. If that's a decent-sized city, then you have opportunities to meet people. The issue is how to narrow those opportunities to find something you will actually enjoy. And if you live in a rural or less populated area and for spice you have to rely on friends cajoling their out-of-town cousins to come stay, then social apps connect you with people you might live close to but not run into otherwise.

As long as you use them with care, dating apps are a great tool. The Pew Research Center found that the number of people ages eighteen to twenty-four who use them has nearly tripled from 10 percent in 2013 to 27 percent in 2016. Ultimately, the stats are in your favor. Use apps regularly, and you'll have many more opportunities to meet people than you would have if you just went to a bar or a club three times a week.

So let's be clear what using apps carefully means. It's why I like the analogy of learning to drive. If you got into a car for the first time, hit the accelerator, and sped off with no attention to what others on the road were doing, you would end up in a hospital. With apps, you also need to figure out where you are hoping to get to, signal clearly ahead of time, observe other people's signals, and, if in doubt, slow down, pull to the side, or stop.

A global authority on digital behavior, the cyberpsychologist Mary Aiken warns that people can easily get into trouble when they mistake the internet for real life. "Cyberspace is a space where we go—it's an immersive environment, not a transactional medium like watching the television," she explains. "When you go online, you're psychologically immersed. You feel anonymous. You are disinhibited."

Sometimes the power of anonymity online can be a good thing in terms of exploring yourself and your sexuality. "No one gets pregnant or an STD from sexting, for example," she says. The problem, however, is that the profile we create online—whether on Match.com, OkCupid, Facebook, or LinkedIn for that matter—is an aspirational self. In the dating world, this applies to both you and a potential date. "So he is presenting this aspirational and highly manipulated entity," Aiken says. "And so are you."

The profile we create online—whether on Match.com, OkCupid, Facebook, or LinkedIn for that matter—is an aspirational self. In the dating world, this applies to both you and a potential date.

In short, both you and the person you're flirting with online are not presenting your real selves. "In cyberpsychology, we refer to Walther's Theory of Hyperpersonal Interaction, which says that when you're in a computer-mediated communicative environment, you only get curated pieces of information about another person, so you have the tendency to fill in the blanks with positive attributes," Aiken explains. "That leads you to idealize this person who you are communicating with."

STRANGER DANGER IS FOR ADULTS, TOO

Add to that what Aiken calls the "stranger on the train syndrome" in which talking to someone you don't know—and might never see again—can lead you to easily disclose personal information. "Online conversations can quickly escalate," Aiken says. "And people can discuss sexualized content much more rapidly than they might do in a real-world context."

Studies on self-disclosure find that people reveal much less to a person who they meet in real life. Online, Aiken says, that level of self-disclosure doubles.

Aiken thinks the internet should come with a "stranger danger" warning for everyone who logs on. "The person you meet online, no matter how well you think you know him or her, is still a stranger," she says. "And much of what you think about that person is idealized—you fill in the blanks with what you want him to be. So you're creating this person." Her point is important: "Online dating is very crowded," she says. "There are four people in it: two real, normal selves, and two virtual selves."

After you've flirted heavily online, it's easy to segue to a real-life meeting with diverging expectations. What happens online is not the same as foreplay IRL. This dissonance between the online/real-world relationship could help explain the sixfold increase in sexual assault associated with online dating.

Yes, you read that right. Sixfold. The most recent statistics come from the UK's National Crime Agency, which discovered what the agency calls a "new type of sex offender"—offenders who were not known to police and had no previous history of assault. Some 71 percent of those reported assaults took place on the first date and in the victim's or the offender's home. "It could be that escalation and amplification of the relationship—moving to talk about sexualized content very quickly—meant that a cyber-intimacy was built," Aiken explains. "So by the time that real-world date takes place, there was an expectation that something more was going to happen."

This is precisely the reason that it is so important to have guidelines (see Rule #5) around online dating—back to my point about driver's ed—and to keep to them. And the first of these rules is obvious. Never agree to meet a stranger—even if you have exchanged thousands of texts and you have dozens of Facebook friends in common—at his or her home for a first date. And if he insists, cut bait. Meet in a public place—a café, a bar—with other people around who informally police you. And always tell a friend where you are going and who with. I have one friend who, when she knows she wants to hook up with a Tinder date, asks to see his driver's license, photographs it, and then texts the pic to a friend telling her what's happening.

Another rule is by all means have a drink, but don't get drunk with strangers, no matter how many secrets you shared online. Approximately one half of all sexual assaults involve alcohol, according to the National Institute on Alcohol Abuse and Alcoholism. (See Rule #8 for more on the influence of alcohol on dating.)

Most women have their own rules around online dating. Cindy Gallop—the founder and CEO of MakeLoveNotPorn, a user-generated, crowdsourced sex-video platform that wants to disrupt the porn industry (see Rule #10)—loves internet dating and has a three-rule process before she goes on an actual date with anyone she has met online.

"It's the Cindy Gallop three-step filter," she says. First and foremost, she chooses men based on their looks—specifically, her own definition of attraction and not what her friends may think if she shows up at a party with him. "I'm not looking for the chiseled jaw, the clean-cut good looks. I don't care about conventional attractiveness," she says. "I genuinely only care about whether he's attractive to me."

Then she tests that by asking for at least three photos, one that is recent and full bodied. "There is nothing wrong with saying this, because everybody knows why you're asking," she says. "If anybody objects, off the list instantly."

Her second filter is a writing test. "I prefer email to messaging, which is a much more casual form of communication," she says. "And I urge women to cut men some slack because they have no idea how to approach a woman online, or off." Her point is, the emails might be clumsy or even crass—but they're a good barometer of whether you sense any potential chemistry. "That's entirely subjective," she says. "For me, any misspellings means it's not happening."

The third and final hoop is speaking *on the phone.* "Young men particularly resist this one," she says. "But insist on it. Ask for his phone number and call him. This is where you will find out whether you like the sound of his voice, which can tell you a lot. You also find out whether you can maintain a conversation on the phone, because if not, then you sure as hell can't over a cup of coffee or a drink."

Ah, the phone issue. I get why people hate talking on the phone. It's tricky. It puts you on the spot. You talk over each other, you get cut off, they call back, and you can't talk right now. You make a joke, but they think you're serious. It's *so* much easier to send a text or email.

But the reason you must call *all* online matches before you commit to meet in person is that a phone call will give you so much more information than any text possibly can. You need to mine for aural clues. There's the sound of his voice for one. On a very basic level, do you like it? Could you listen to him for an hour? Does he sound the age he says he is? Does he sound kind? Dozy? Sharp? Psycho?

Have a list of light questions ready. Kick off with a reference to a local sports team, a band in town, a book you're reading, a news event.

The goal is to figure out if you speak the same language. I don't mean literally, but do you get each other's references and cues? Is he polite or aggressive? You need a sense of his values. Does he care about the things you do? Forewarned is forearmed. And remember, he is also listening for clues. Do *you* sound like yourself?

The goal is to figure out if you speak the same language. I don't mean literally, but do you get each other's references and cues?

This is not just about protecting your valuable time. It's about using your psychological detective skills to preserve your emotional energy. If you find yourself on a never-ending carousel of internet dates, something's not working and you need to hone your selection process.

Think of it as you would approach hiring someone at work. Yes, the HR department scours LinkedIn, but they call, Face-Time, or Skype before engaging a manager's time to meet a candidate. It's exhausting to keep swiping, getting your hopes up, and then meeting people who don't work out.

Gallop loves the efficiency of the phone. "If it's not going well, you are able to say at any point, 'It's been nice talking to you, but I don't think it's going to work out. Bye.'" she says.

However, if you have a nice conversation and like the sound of his (or her) voice, then set up the first date. Gallop guarantees it will go fine because this person is now no longer a total stranger. "I've met with men with whom there was absolutely no chemistry, but because of those three filters, it was a perfectly pleasant first encounter as opposed to an awkward, embarrassing one, which none of us have time for," she explains.

Aiken agrees a phone call first is important. But her bigger concern for women looking for love online is just how male-dominated dating apps are. "We have spent so much time fighting for our rights as women, and yet we live in a cyberspace that is almost exclusively designed by men," she says. "Women need to be actively engaged in helping to design it."

Thankfully, that is beginning to happen. Tinder and Match are among the top ten most used dating apps, according to a study by Applause, an app analytics company, but two of the hottest apps lately—Bumble and Coffee Meets Bagel—were founded and are run by women. Whitney Wolfe left Tinder, which she cofounded, and set out to create a "positivity social network" for young women called Merci, rooted in kindness. The tagline was "compliments are contagious," which meant users could leave only compliments on one another's pages. "I wanted to do something in response to the lack of online accountability," she explains. "When I was a kid, you couldn't bully someone in the classroom without consequences. You can't run a red light or speed without repercussion. Where are the consequences digitally? They don't exist."

When her business partner asked, "Why not do this in the dating world? It needs this," Wolfe, at twenty-eight, looked at her own experiences as a young, single woman and realized he was right. She started Bumble because she felt that she had succeeded in so many aspects of her life—except dating. "I was confident enough to travel the world, study abroad, and start my own company," she says. "Yet, I always felt very disempowered when it came to dating. It all came down to me not being able to make the first move," she explains, recalling "hours of agony" in college, waiting for guys to text after meeting them. "Whenever I did make the first move, I was shunned or guilted for it by friends or by society's standards," she says. "That dynamic is just insane."

She started asking all her friends about their dating experiences and realized that she was not alone. "Almost all of the women I know had been in dysfunctional relationships," she says. "All these wonderful, brilliant women, in this constant state of disarray surrounding dating—it really comes down to power and gender dynamics." What began as a social network morphed into Bumble, a dating site that placed women in the driver's seat, a move Wolfe sees as the antidote to unhealthy relationships. "It lets women make the first move with no judgment, stigma, shame, guilt, or blame," she says.

To ease the awkwardness of women making that first move, Bumble—which brilliantly markets itself as a dating app for people who would never use a dating app—has a set time limit. After the initial prompt, the match disappears in twenty-four hours. Empowering women to make the first move, Wolfe adds, has been truly exciting, and to Aiken's point, a move toward correcting the gender power imbalance. Clearly women agree: The company celebrated 250 million first moves made by women on

International Women's Day 2017. "If we keep doing what we're doing, perhaps we can rewire and reconfigure the dating world," Wolfe says. She'd like to start with a misconception that men are all lecherous monsters. "People remark time and time again, 'Oh my gosh, the guys on Bumble are so handsome, high-quality, educated, and smart,'" she says. "My response is, 'You'd be surprised by how many men want a confident woman who has her own voice.'"

Use your voice and your best-suited app (see the next rule!) to help you find the partner who makes the most sense for you.

CASE STUDIES

Megann, 25, on how internet dating saved her small-town love life.

Megann grew up in a rural town outside Detroit. By the time she graduated from high school, she had dated anyone who was, in her words, even a remote possibility. "I like the more sensitive, soulful types and grew up in a place where they were rare," she explains. She went to a local community college, and her options did not improve. "It felt like high school all over again," she explains. "You pretty much had to hope someone brought a friend from out of town to a party. And that still felt like needle-in-a-haystack odds."

She started playing around with online dating, first Tinder then Plenty of Fish (POF) and Swoon out of "pure boredom." "I never thought anything would actually come from it besides maybe talking to some fun people," she says. She flirted online with a bunch of guys who seemed interesting and wound up meeting a few in person, but none were a real match. "I still felt lucky," she says. "They were all good experiences, just no one I wanted to actually date."

But then Drew got in touch with her through Zoosk, which she liked because it was "easy to use, and free." After a witty back-and-forth that lasted two hours, he sent her the message: "What would it take to go on a date?" To which Megann responded, "Just ask!"

He did. She agreed to meet him for lunch at a nearby Chili's—and was not ready to say goodbye after he paid the bill. "We went up to a USS warship that is parked nearby," she says. The date lasted twelve hours. "Unlike the other guys I had met, he lived up to my idea of what he would be like," she says. "Super-bright and motivated. I was inspired."

Drew had just come out of the military, so he was also well traveled. "I found that incredibly attractive," she says. "Most of the guys I grew up with had no aspirations to visit Detroit, let alone leave the country." He had been living with his mom for two weeks, in a town an hour and a half away. "We never would have met if it weren't for the internet," she says.

During that first date, they covered lots of territory—and discovered a mutual love of comic books, *Star Wars* movies, and Mexican food. It took a few more dates before Megann mentioned something she wanted him to know before they got too serious. "I have always known that I don't want kids," she said. "So I was relieved when I told Drew that, and he smiled and said, 'That's cool. Me neither.'"

Now, four years later, they live together, with two cats, and are saving up for a trip to Europe, Megann's first trip outside the US. She is certain that without the internet, she would still be single. "Growing up in a small town like mine, we had such limited options," she says. "Plus, I was not a social person, so I would have been screwed."

Kelly, 36, on her "dating-palooza," or 500 dates in less than two years.

A year and a half into her marriage, Kelly's husband said he wanted out. Kelly was thirty-four years old and went into therapy to figure out what went wrong, as well as what role she may have played in it. "I really thought I had married my teammate for life," she says. "But as soon as we married, it started to unravel. I knew I had to honor the grief and work through all the emotions before I even thought about dating again. It wouldn't be fair to displace any unresolved emotions on another person."

Kelly did a dating detox for several months before she began what she calls her social experiment: "dating-palooza." The founder of Shynebyte, a New York–based talent strategy company, she applied similar strategies to finding a partner that she does to matching the best client with the right job. "The world of recruiting is complex," she says. "So I created a five-A model to optimize the process for my work, which I applied to my dating approach: Align, Attract, Assess, Acquire, and Acclimate." Align, for instance, is to proactively define what you want in a partner, whereas Assess involves accepting that dating is messy and so being open to all potential outcomes, including making male friends or business connections. She signed up for Tinder first, chose six photos that she felt represented the different dimensions of who she was, and wrote a bulleted bio that emphasized her quirkiness, such as her goal to visit each Beach Boys "Kokomo" location (she's halfway there).

Kelly then committed thirty minutes a day to swiping, and she experienced both swift requests for her phone number and drawn-out messaging that never graduated to texting or a phone call. She preferred the former.

In the year and a half that followed, she went out with 351 men and on at least five hundred dates, often averaging two a night. "It was exhausting," she says. "Even thinking about it I get tired." Those nights when she was meeting more than one man could also get hairy: "I did get caught once running late to the next date, whom I had planned to meet across the street," she says. "I admitted why I had to run—it didn't go over very well."

Her attitude about meeting people remains open. "I've actually set people up with my friends," she says. "I see this as a value-add ecosystem." She has had bad dates and great dates—and has met some really interesting people, as well as a few creeps. "One guy wanted to play truth or dare over text," she says. "I found it a refreshing break from 'Hey, what's up?'" She chose truth, and the two went back and forth with questions she appreciated, like "What scares you most?" But then, out of the blue, he texted, "Spit or swallow?" "I blocked him immediately," she says.

That was not the only man she has ever blocked. Another guy texted her, "Something tells me you've been a naughty girl, and need to be punished," she recalls. Kelly replied with, "Your intuition sucks." "I could make a coffee-table book of all the unsolicited dick pics I get," she says. The new thing, she adds, are "cumming videos." "Men I have never met, jerking off in a video, calling my name as they cum," she says. "It's very disturbing."

She has also gone on multiple dates with men she thought had potential. "I went on four dates with a guy I liked a lot," she says. "But then, he said, 'When the time comes, I hate condoms.'" She told him they were necessary for her, no matter what, and his response floored her. "He said defensively, 'Do you think I would sleep with a girl who's not clean?'" she recalls. That was the end of that relationship.

Another guy she dated was in the tech world, so when she used the term "elastic load balancing" in their conversation, he got quiet for a moment. He then said, "I think you are too smart for me." At first she thought he was kidding, but his face remained stoic. The date ended there. "I was shocked and slightly embarrassed until I realized, 'Fuck that! I'm not ashamed that I know about the cloud,'" she says. "I'd only want to be with someone who embraces my intelligence."

Shortly after a few underwhelming dates, Kelly turned her focus to Bumble and The League. She also adjusted her parameters to include men in their early forties, as her theory is that they are less intimidated by smart and successful women. So far, that move has been positive. "Recently, I had a date with a guy who asked if I date a lot. I told him I had met 350 dates," she says. "And he said, 'What is it about you that you felt the need to date 350 people?'" The question led to a second date.

Rule #5

Choose the right recipes for your dating type.

The good news is that online dating has come a long way since its early days, when your choices were basically Match.com or JDate, and you felt embarrassed admitting to friends or family that you were trying them. With our growing expectation that the answers to any of life's problems can be found in our phone, online-dating options have proliferated, as have the number of people using them. As previously stated, according to a 2015 study by the Pew Research Center, 15 percent of American adults have dated online, and participation by eighteen- to twenty-four-year-olds has almost tripled since 2013. And according to the same study, people over fifty are one of the fastest-growing segments of online dating, so it's never too late.

The sheer number of options and vastness of opportunity may make embarking on online dating overwhelming, so I have broken the process into steps with tips for each phase. This is the longest rule in this book for a reason: It is a guide to help you navigate all that is out there. Yes, apps are tools, but some are better suited for you, so choose well and wisely. There's no shame in this game!

STEP 1: CHOOSE AN APP THAT'S RIGHT FOR YOU

Deciding what dating app or website to use can be daunting. You don't have to stick to one, either; it's okay to mix it up, though you don't want your profile out there on every single site. What makes sense is to choose a lead one that most clearly answers your current needs. At *Cosmo*, readers love to take quizzes. I thought one here might make the process of choosing an appropriate app a little less hit-and-miss and a lot more fun.

And whenever I think of conversations about dating, I think of Carrie, Samantha, Charlotte, and Miranda lingering over brunch and dissecting their love lives. I loved their frank attitudes about sex and varying levels of interest in intimacy and dating, totally independent of their age. And millennials do, too; *Sex and the City* is on perennial repeat on E! as it clearly still resonates.

So I am borrowing their attitudes for this conversation here.

Samantha's unending appetite for fun, Carrie's musings on the universality of desire, Charlotte's faith in postdivorce love, and Miranda's clear-eyed return to dating after having a baby . . .

See which woman's attitude best aligns with yours in order to discover which dating app is likely to work best for you. But do try several before locking into one. You never know what might—ahem—click for you.

Quiz
1. *What's your attitude toward dating?*   A. "I'm a trisexual, I'll try anything once."

B. "Being single . . . means you're pretty sexy and taking your time."

C. "I've been dating since I was fifteen. I'm exhausted, where is he?!"

D. "It's all about timing, you gotta get 'em when their light's on."

2. *Tell us about the pool of dating prospects you're willing to dive into.*

A. This swimsuit has easy-to-undo ties for a reason.

B. There are so many fascinating people in my city, and I know most of them. Why not make use of that?

C. He has to be well-bred, handsome—a doctor would be wonderful, especially a surgeon—tall, smart, no mommy issues, doting. It would be nice if he liked sex . . .

D. I have no tolerance for fools or people who will waste my time.

3. *Are you willing to pay for a dating app?*

A. Why pay when you can play for free?

B. I spend money on shoes, not dating. And hello, I'm a freelance writer. The few dollars I do have are going to Net-a-Porter, not to net a man.

C. Of course. I want to find the best, and I'll pay more to find my perfect match.

D. Yep. I'm happy to pay for any good service. I basically need an electronic version of my housekeeper/nanny Magda for my dating life.

4. *What are your deal breakers?*

A. Honey, deal breakers are for the uncreative. I'm a stud maker . . .
B. I wonder . . . in today's connected world, why would anybody think sending a dick pic is a good idea?
C. My heart is open, and I want someone who is as ready to find his soul mate as I am. No players, please! (Also, I'd really prefer a guy who puts his boxers in the laundry basket—not on the floor where he took them off.)
D. Unambitious or uninteresting men need not apply.

Mostly A's: Samantha Jones

You've got a carefree attitude about sex, love, and dating, like the ever-sassy Samantha. Apps with a wide range of choices that have no fee will give you your most fabulous online-dating experience. It's not that you're *not* open to a new relationship; you're just down to clown along the way.

TRY: Tinder, Happn, POF

Mostly *B*'s: Carrie Bradshaw

You're a romantic at heart, but you're certainly not opposed to taking a "lovahhh" or two along the way. You trust your own instincts when it comes to shopping for anything, be it men or Manolos. Apps that put you in the driver's seat and tap into your social network are going to be big when looking for a mister.

TRY: Bumble, Coffee Meets Bagel, Hinge

Mostly *C*'s: Charlotte York

Your Pinterest boards for wedding dresses and East End mansions are renowned. You know what you want; you're not going to settle for less, and if finding the best costs money, so be it. Apps or sites that offer serious suitors with similar interests will help you find the baldy of your dreams.

TRY: eHarmony, The League, Twindog (if you need someone who loves your dog the way Charlotte loves her King Charles spaniel, Elizabeth Taylor)

Mostly *D*'s: Miranda Hobbes

You've worked hard, and you know the world. You're determined that this time, you'll use your resources to make dating a more logical and smooth experience. Who

has time to swipe for all eternity? You want results, and
you want them fast, and you want algorithms to back
them up. Apps and sites that weed out the slackers and
the stupids will push you in the right direction.

TRY: OkCupid A-list, Match.com, EliteSingles

If wisdom inspired by *Sex and the City* doesn't turn you on,
heed these words of wisdom from online-dating expert and digital matchmaker Julie Spira:

1. Be honest about what you're looking for. If you know you
   want to get married NOW and have children, don't waste
   your time on an app that's known for more casual relationships.
2. Don't become victim to what I call "Online Dating Fatigue."
   I recommend using two to three apps at a time, plus a desk-
   top version of a dating site. Any more than that and you'll be
   overwhelmed and frustrated. Plus, you don't want to go on a
   date and ask the guy about Tinder, when you met on Hinge!
3. Wait at least three weeks before upgrading to premium pay
   options. On most apps, the newer you are on a site, the higher
   and more frequently your profile will appear. After that new-
   bie glow wears off, boost your search with premiums. For
   example, Tinder Plus lets you undo mistake swipes, change
   your matching location (handy if you're planning a trip), and
   see who has liked you in the past. Bumble Boost has similar
   features and also extends the amount of time you have to
   engage a match.

STEP 2: MAKE YOUR PROFILE

Now that you've chosen the app, it's time to get into the game. The most important weapon in your arsenal is your profile. Think of it this way: your profile is to dating as your résumé is to your job search. It is your chance to market yourself and put your best foot forward. It is the one space in the dating process where you can completely control the messaging. And while it might not be easy—most women I know still have a hard time bragging about themselves—try to have a bit of fun with it.

Think of it this way: your profile is to dating as your résumé is to your job search.

When one wants to launch any new creative endeavor, the first step is always due diligence. And there's no reason dating should be any different.

Get on whatever site or app you've chosen and see how other women are portraying themselves and putting themselves out there. Look at their profiles from the perspective of someone who is looking to date you, as well as them. What catches your eye about their profiles?

Keep an eye out for the mix of photos, how detailed the bios are, the tone and overall vibe, and what feels most—and this is important—*you*. Ask yourself whom you see a kinship with. Whom would you set up with a friend? Screenshot the ones you find most and least appealing. Take a look at the ones you like and the ones you don't and see what they have in common with each other. Do you like the funny ones? Are you turned off by people who seem as though they're trying too hard? Use the ones you like as inspiration not for *what* to say but for the tonal mix of your profile.

Do you like 25 percent funny, 10 percent cocky, 65 percent earnest? Or are you more of a 50 percent sarcastic, 50 percent brass tacks kind of girl?

Determining the answer to questions like these is where your dating board of directors comes in handy. At a certain point in their careers, many successful women I know rely more heavily on their personal "board of directors"—a group of people they trust who have varying expertise and insights—for advice than on any one mentor. My friend Carolyn Everson, who runs global marketing at Facebook, first introduced me to the idea, and I think it's powerful. I see no reason not to adapt the concept to your dating life. Think about the people who know you best and genuinely have your best interests at heart. These are people who care about your happiness and whom you can use as a sounding board or an occasional comforting shoulder. Do not, however, turn them into a chorus in a Greek tragedy of indecision. You're not screenshotting every person in your contacts to see if you should swipe right or left.

Creating your profile is an ideal time to call on your board of directors. Email them and say, "Quick: give me the one line you would use to tell someone what you love about me." Use what they say to get you started and set the vibe. If everybody writes back and says, "You're the friend who is always there for me," then you might want to consider writing a profile that reflects what a good friend you are rather than something more tongue in cheek. Also, keep in mind who and what you're trying to attract. Remember what app you're on. If you're on an app to have fun, make your profile more fun. If you're on an app looking for a long-term relationship, make your profile more sincere and direct.

While you want your profile to be good, don't obsess over it! Get something out there. Done is better than perfect here. This is a malleable product. When new companies launch, they go to market with something called an MVP: a minimum viable product. It's the most boiled-down version of what your end product will be. It's not perfect, but it will get things going so you can see what works.

Personally, I would want Amanda Bradford on my board of directors. The Stanford business school graduate created the dating app The League to address her own frustration with the lack of quality on most apps.

I asked her for her profile dos and don'ts:

DO: Show one to two full-body shots and clear face photos.

DON'T: Include pics of your hottest friend. Get her out of your photos. For once it's okay to make this about you.

DO: Show off what makes you unique: Do you play squash or lacrosse? Are you the festival type looking for someone who can enjoy Coachella with you? Do you knit, build drones, deliver Meals on Wheels on the weekend? These are all great things to feature in at least one of your photos. While you may not get hearted by everyone, if you're trying to optimize for someone who has similar interests, this is the way to do it.

DON'T: Post old pics. Keep your pictures current. We were all kids. We were all cute as kids. Then you grow up. Show that in your picture selection. It's nice that you were that skinny when you graduated from college, but anything older than five years is false advertising.

DO: Wear white. The top-performing women of The

League have at least one photo in a white dress or shirt. Weird, right? So yes, do an eye roll at the historical symbolism of white dresses as purity and innocence, but it is true that everyone does look great in white.

STEP 3: GET GOING

Think of going on dating apps as January gym behavior, which is fitting because January is also the busiest time of year for online dating, as it kicks off the postholiday cuffing season (the time of year when people look to get into relationships, as they're more eager to be paired up between October and February when it's cold, and relatives ask demanding questions from Thanksgiving to Valentine's Day). It's all well and good to join a gym in January, but if you don't push yourself to go as much as possible in that first month, you won't make it a habit and get into the swing of things. Start working that online-dating muscle pronto—and stick with it! If the gym analogy doesn't work for you, think of this as a cocktail party where you made the effort to put on lipstick and brush your hair. Since you are already there, you might as well make use of it and talk to people.

On that note, there are different schools of thought on whether you should sit back and wait to hear from potential suitors or take control and initiate messaging with potential partners who catch your eye. It won't surprise you here that I'm all for leaning in. When there's a pair of Altuzarra boots I want on sale at Barneys, I'm not waiting for them to ask me to buy them. I'm snagging them before they're gone. The same theory holds here.

Serendipity is one of the most frustrating—yet eminently watchable due to how charming the actors are—mainstays of basic cable. Two unmarried people come across someone with whom they have an immediate intellectual and chemical connection. Instead of capitalizing on that and saying, "Well, this doesn't happen every day," they put their romance in the hands of fate, effectively saying that if it's meant to be, the world will bring them back together. Absurd. The world does not have your back to that degree. So when you feel a connection and see something you want, take steps toward it. Don't sit back and put your trust in the universe.

If you do reach out first, you will need an opening line. It's easiest to have a go-to so you don't have to overthink it every time.

Dating apps are so pop-culturally pervasive that they've become a common theme in everything from TV shows to songs. When Aziz Ansari's Netflix show *Master of None*—his delightful modern comedy of manners—debuted its second season, everyone was talking about his character Dev's clever opening line on dating apps: "Going to Whole Foods, want me to pick you up anything?" It caught on because it works; it's charming and specific, and it speaks to Dev's funny, foodie personality.

So cook up your own signature line with room for personalization that will show your potential match that you gave their profile at least a cursory read. If you can't think of anything, one young editor I work with who's an online-dating veteran brainstormed a bunch of fun and flirty possibilities:

"So, should we have 🍺 and 🍸 or ☕ or 🍵 on our first date?"

"Before we get into it, I have to know if you're one of those people who clap on a plane when it touches down on the runway."

"Quick: what was your first concert? No shame. Unless it's Nickelback, then shame."

Or if all those feel off to you, go with tried, true, and simple.

"Hi there . . . how's this all going for you?" or "Hi there . . . love your photos. Tell me about the one you almost posted but nixed."

The point is to get a conversation going and see if there's a click. Many of my friends say that after three or four exchanges, they switch to email or text.

Do what feels comfortable for you.

STEP 4: MEET IRL

At some point you will need to meet face-to-face. This may seem as though it should go without saying, but I'm doing so because it is so easy to fall into the trap of swiping and messaging and not actually acting on any possibilities.

Going back to the earlier gym analogy, when you buy a membership or re-up the ClassPass, you think about how many classes or sessions you'll aim for each week. The same applies to dating.

Give yourself a goal for how many dates you'd like to have each week. It's a numbers game, after all. The more people you meet, the more opportunities you have to find someone you actually can spend time with. Or hell, someone you would be thrilled to be with.

So just like your SoulCycle resolution, make an achievable goal for yourself rather than one that is so ambitious it becomes daunting. Twice a week seems rational to me, but do what works for your schedule. And keep in mind, according to current dating practices, these don't need to be dinners. They can be coffees, drinks, or strolls through a bookstore. And while a first date at the local cemetery can have a certain spooky charm, I'm thinking it's not the best idea. The what or where is not so important, it's the actual doing that matters.

We've established that I think it's essential to talk to your prospective date on the phone before meeting in person. But I realize that that might feel impossible for some people (millennials, I am talking to you!), so here's an exercise to help you determine what's right for you:

Make a list of the last three people you spoke to on the phone for more than thirty seconds. If it was your best friend ten times in the last twenty-four hours, your mom three times, and your work spouse for a quick check-in, then you clearly enjoy phone time and should have some with anyone you're considering dating.

If that list is limited to the call you made to thank your aunt for the macramé pillow she sent for your birthday and the rant at your dry cleaner for destroying your leather pants, you're probably not a phone person, so making this a hard-and-fast rule will only hold you back.

Either way, the journey from connecting online to meeting in person should not take more than a couple of weeks, or it probably wasn't meant to be.

Horrified by the idea of the date itself?

Here's my take on best practices for your first encounters . . .

What should I wear?

Something you feel confident, attractive, comfortable, and, most important, *you* in.

Do not wear the thing that fit you last year and will fit you again when you lose those eight pounds.

Do not wear the thing that you think shows your secret inner sex kitten.

And probably, do not wear shoes you can't walk in.

What if I can't find him in a crowded bar? How do we link? WTF am I doing?

Relax.

You have the benefit of dating in the age of cell phones. Exchange numbers before the date if it wasn't how you were already communicating. And remember that you both know what each other looks like. As you are looking for him or her, he or she will be looking for you.

What if I run out of things to talk about? What if there are awkward silences?

It's not weird to come ready with a couple of preplanned topics. It's an oft-quoted adage to read the latest news. In my day, that meant the front page, editorial page, and style section of the *New York Times* or *Washington Post*. Catch up on current news in your favorite format before any potentially awkward social situation so you're full of interesting chitchat.

If news isn't your thing, that's fine. But come in with something current that's interesting to you to discuss—a movie you've just seen, a book you've just read, or a new app you've just tried.

What should I do to impress him?

Wrong question. If you're thinking about impressing him, you're not being yourself and, likely, you're not paying attention to what you actually think of him. Or her. Or them. Whatever, the point is to be the best but not an insincere version of you. Don't say you like to hike on the weekend if the outdoors and exercise make you cringe. You're there to see if you like him, not to win over a new conquest. Remember to check in with yourself mentally during the date to determine how hanging out with him is making you feel.

But where should we actually go?
If it's up to you to pick, think of where you'd want to meet someone you admire. That will force you to choose someplace classy with comfortable decibel levels.

Best rule of thumb: unless he insists otherwise, make the plan for a drink and say you have a dinner scheduled after. Bethenny Frankel once advised a friend of mine that even if you're basically the heart-eyed emoji after an hour, don't cancel your imaginary dinner. Leave him wanting more (also, it wouldn't do to have him think you're a flake).

Go on the dates! Have the fun. Don't get too drunk (see Rule #8). And really, just be in the moment.

STEP 5: THE AFTERMATH

You've gone on the date. Well done.

My general point of view is that unless the guy or girl was a complete horror show, it's polite to send a quick message the next day. Don't worry, you're not leading them on if you say, "Thanks so much for last night, great meeting you." That pretty much translates to "Bye, best of luck."

For more detailed advice on the next-day follow-up, I turn to another one of my board of directors, *Cosmo* dating expert and brilliant writer Logan Hill.

Dos and Don'ts for the Follow-Up:

DO: Keep it simple. A basic "had a great time, let's grab dinner soon" is plenty and lets him know you had fun and want a repeat. A little inside joke inspired by your date is even better.

DON'T: Stress out and rewrite your text a thousand times over. It's just one text message. You probably send hundreds of texts a day, so don't overthink it. The only mistake you can really make is to come on too strong. Don't cluster-bomb him with Bitmojis and GIFs right off the bat.

DO: Forget the old rules. There's no twenty-four-hour rule or forty-eight-hour rule to following up. And if you live by such a rule, the odds are that the guy won't know anything about it, and you'll be playing a game by yourself. When you respond after a date depends entirely on what feels comfortable to you. There are no "four words" you can say that will make him melt—and no "three secrets" to dating. So text him that night if that's your style. Text him in the morning if you're a morning-after person. Or never text him at all if you want him to take the lead. It's entirely up to you.

DON'T: Force it if you don't feel it. As I've mentioned earlier, a first date is not the perfect test, and many second dates go on to be unexpected successes. But if you're *really* not into a guy after the first date, there's no obligation to make it work. You don't owe each other anything. Move on.

DO: Embrace your own desires. I think the first, second, and last thing a dating woman should be thinking about is what she wants. This is a time of unprecedented choice—and the options can feel overwhelming. But it's also an incredible moment of opportunity. Embracing that opportunity means imagining, out of all those many options, the type of relationship that would work for you and maybe even thrill you.

There are no "four words" you can say that
will make him melt—and no "three secrets" to
dating. So text him that night if that's your style.
Text him in the morning if you're a morning-
after person. Or never text him at all if you want
him to take the lead. It's entirely up to you.

DON'T: Play it too cute. If you prefer to be chased, then
by all means let him wonder. Otherwise, tell him what you
want—or, better yet, suggest something specific. The days when
men had to make the first move and women waited by the phone
are over. We don't want to go back to that, do we? It's scary
sometimes to ask for what you want, but the odds of getting it if
you do are just so much higher.

WHAT TO DO IF YOU RUN INTO THESE PEOPLE ON DATING APPS

YOUR FRIEND: Green Light. If it's your friend, there's no harm
in swiping right—or liking the profile—and having a laugh together about how weird or awkward it is. Worst case? He doesn't
swipe right, but he still won't know that you did.

A COWORKER: Yellow Light. As everyone knows, dating
someone you work with can be a slippery slope. Before swiping,
think about the consequences. If things go to hell, would showing up to work and seeing him or her every day be torture? If
you decide that it's worth it, say something like "So, this is what
you do when you're not forecasting revenue?"

AN EX: Red Light. Unless you're trying to reopen the relationship or reexamine why you broke up, rekindling a burnt-out flame probably isn't the best idea.

A FWB: Green Light. Well, you're probably both on the app for the same reason . . . either looking for an additional FWB or maybe just a "with benefits" situation. So unless you've developed feelings for each other (which is a pretty big no-no in FWBs), then you should be good.

WHAT IF HE TURNS OUT TO BE THE WORST?

The least bad scenario is that he's just not nice.

The worst: he actually does something aggressive or makes you feel unsafe. How do you handle it?

If he sends you a dick pic or cum video, block him immediately and don't send anything back. If you're not into penises flying through cyberspace and onto your screen, cut off all contact. You also don't want any of that appearing on your screen at work. And it goes without saying, you should have your own personal account for all this. Never even dabble with online dating on your work email account! I remember a colleague handing me his phone to check out a photo he had taken for a work reference just as a match landed from Tinder. I didn't need to see it, and he didn't need me seeing it, either. Awkward!

If he sends you inappropriate messages or is angry about you blowing him off, *don't engage, don't engage, don't engage.* If it gets uncomfortable or out of hand, Julie Spira suggests reporting him to the site as abusive.

The absolute worst nightmare is if things turn violent on

a date. For that, I asked law enforcement veteran Steve Kardian for his tips, which he fleshes out in his book *The New Super Power for Women: Trust Your Intuition, Predict Dangerous Situations, and Defend Yourself from the Unthinkable.*

To start, Kardian urges women to remember that when you're dating online, the only thing you can be sure of about the person you're communicating with is that he has a computer. So how do you know he is who he says he is? Kardian has tips:

USE ONLY REPUTABLE SITES: "Not Craigslist," Kardian says. "But those where they at least ask you to register through your Facebook account so there is a digital footprint." This helps weed out scammers and catphishing, where people look for ways to get your personal information (identity theft) or bank information (fraud). On that note, Kardian says to never send anyone personal information—especially your home address—until you are sure the person is trustworthy.

TAKE IT SLOW: Ask questions that give you ways to verify his identity. Beyond his name, where does he work? What kind of car does he drive? Does he have a pet? "We have a front stage and a backstage," Kardian says. "When you meet someone, usually that person is on the front stage." You want to make sure his backstage is not criminal or creepy.

RUN A GOOGLE NAME SEARCH: If he comes up on five or six dating sites, that is problematic. If he does not have a digital footprint, that is also a red flag.

DO NOT SEND SEXY PICS OR ANY PERSONAL INFO: You really don't know this person until you have spent real time with him. So don't send any sexy photos or even partake in a sexy Snapchat until you feel you can trust him—and that takes many dates. "We are seeing a huge increase in sextortion cases," Kardian says. "This is all based upon the growth of social media and the internet."

MAKE A SAFETY BLUEPRINT BEFORE THE FIRST DATE: Kardian agrees that talking on the phone is a good safety step before meeting IRL, though that actual meeting is where you will get your best clues. But even before you do that, he says every woman must make what he calls a "safety blueprint." "Think about all the things that could potentially go wrong, and come up with a response," he says. "So what if you want to leave and he is not letting you? Then what?"

If you think through all the possible scenarios, you are prepared for the worst and can still hope for the best.

Map your safety blueprint.

BEFORE THE DATE: Tell your best friend/mom/sister whom you are going on a date with and where.

DURING THE DATE: Do a gut check. Ask yourself how you feel about this person physically. "Your gut instinct is your best friend," says Kardian. "If you have a little sense of foreboding—the hair on the back of your neck stands up or you get a queasy feeling—trust it."

YOU WANT TO LEAVE, BUT HE'S INSISTING YOU STAY FOR ANOTHER DRINK: Tell him that your best friend is meeting you at the bar/café/park in thirty minutes. Say that your mom had surgery and you need to check in with her. Say that you're not feeling well or that you have your period and need to go to the restroom. And then just leave, or call the friend and tell her to come meet you.

HE WANTS TO TAKE YOU SOMEWHERE ELSE: Kardian suggests taking a photo of him and texting it to a friend. You can even joke, "If something bad happens to me, my friend knows I'm with you." A good guy will think you are smart. A bad guy will, too, and that alone could protect you from harm.

If he says, "Let's go back to my apartment," and you say, "I'm not ready for that," look for his reaction. "If you get a flash of contempt, anger, or disgust, heed that," Kardian says. "The combination equals big red flags."

If he won't take no for an answer, then know the same could be true when it comes to sex. "Anyone who does not respect 'No' is someone who needs to control you," he adds.

Do not put yourself in a position where you are alone with him. "If a predator wants to do something bad to you, he needs isolation and control," Kardian says. "He has to get you alone— either in his car or back to an apartment."

If you think it's safe to go with him to another place in his car, take a photo of his driver's license or license plate and text it to your friend.

YOU'RE BACK AT HIS HOUSE AND REALIZE THAT IT'S A MISTAKE: "Don't let on that you are scared," Kardian says. Instead, say, "You have beer breath—can you brush your teeth?" Or say, "I'm allergic to your cologne—can you shower?" Or say, "My stomach hurts. I think I'm going to throw up." And be very convincing. These excuses will give you enough time to leave or escape in most situations. If you say, "I have an STD," and he says, "No big deal, so do I!" then you may be dealing with a psycho. Remind him that you sent your friend his photo, and she knows who he is and that you're with him.

"The most important thing you have to do is continue to be persistent," Kardian says. "At that point, resort to self-defense."

A woman's strength is in her lower quadrant, according to Kardian. "Leverage, technique, and timing can be used to great effect, even if he is much bigger than you are."

Beware of revenge porn.

Say you do wind up sleeping with him, and it turns into another date and another. I am never going to tell you to keep the phone out of the bedroom, but keep your phone out of the bedroom until you feel really comfortable with him. Revenge porn, where he blankets your colleagues, parents, and the web with your naked photos, is seriously on the rise and hard to get taken down. You have to be really careful about who you let take naked pictures of you.

If someone is pressuring you to do it and you don't feel comfortable, heed that instinct. And don't see that person again. This is not something to be taken lightly.

Most work employees are told to imagine how they would feel if their emails about the company were published on the front page of the *New York Times*. Think how it would feel if this person posted your naked photo to his Facebook page. And then imagine trying to call Facebook or Google to get them to take it down. Good luck with that. One in twenty-five people are threatened with, or victims of, "nonconsensual image sharing" according to a 2016 study done by the Data & Society Research Institute.

In the way that lawyers always plan for worst-case scenarios, think about what would happen if any of the photos he or she has taken of you went public. Or think about how you would feel if he sent those photos to everyone at work. Would you be able to laugh it off? Or would that make you want to curl up in the fetal position under your desk? Or worse? These are unfortunately the questions you need to ask yourself before you start letting a partner film you doing anything intimate.

The same with sexting.

A study done by McAfee, a company focused on fighting cybercrime, called "Love, Relationships and Technology" found that 50 percent of all adults say they have stored intimate information about themselves on their phones or personal devices that "leaves your reputation at stake," and another 50 percent of all adults say they used their mobile phones to exchange intimate information. I'm not saying don't pose for Picasso, but 16 percent told McAfee they had exchanged intimate information with total strangers.

As the study authors conclude: "We assume intimate exchanges and private data are safe with loved ones. But what if the relationship goes bad? Twenty percent of people say they log into their significant others' Facebook profiles monthly. This may seem like innocent fun, but information posted without your knowledge may cause you embarrassment and harm your reputation."

Unfortunately, it reverberates more on a woman than a man. Justin Garcia, PhD, an evolutionary biologist and sex researcher at the Kinsey Institute at Indiana University, ran a study about sexting and found 73 percent of people felt "discomfort with the unauthorized sharing of sexts beyond the intended people." "So when you send a naked photo to your boyfriend or girlfriend, you might not assume they're going to share it with their friends," he says. When they did share without consent, it was with three or more people. "The ability to digitally share an image rapidly with lots of people is the risk," Garcia says. "It's not taking the picture itself—but the break of trust. And in that way, the internet and technology changes the risk profiles and the expectations involved in our erotic lives."

CASE STUDIES

Marcelle, 53, on how internet dating made her feel "fuckable" in her fifties.

Marcelle started online dating in June 2015 at the age of fifty-one because, she says, "I hadn't had sex with a human being in three years. Something had to give."

A single mom of a teenage girl, Marcelle cofounded the feminist magazine *Bust* in 1993, and now runs BARB, a website for women over forty. The tagline is, "for ladies like us." "It's for girls like us in this next stage of life, where we're dealing with cellulite that has migrated from the back of your thighs where you can't see it to the front," she says. "Or you sneeze and the wee comes out. Like do we have to take a second pair of underwear every time with us now?" But, she adds, "women like us" still want to have sex.

A huge fan of the vibrator, Marcelle uses hers "every day like clockwork." But after three years without "real intercourse with a human being," she was ready to go online, despite her concerns about online dating for older women. "I kept thinking of the Amy Schumer skit where she and Julia Louis-Dreyfus and Tina Fey ship older women off in a boat because they're no longer fuckable," she says. Meanwhile, Marcelle has always kept herself in great shape and felt sexy. So she decided to go for it and put up a picture of herself wearing a push-up bra and tight T-shirt, which many women her age might feel is inappropriate. "I wanted to show my body, like 'Look! I'm hot,'" she says. "You have to put away the crisis in your head that you're not sexy enough."

She signed up for OkCupid, Tinder, and Bumble—and 80 percent of the messages she got were from much younger men who, she deduces, "want to be with a woman who's sexually experienced." If she is in the mood, and her daughter is away at a sleepover, she takes them home. "I'm not romantic about it," she says. "I've always been able to separate sex and love." But she has altered her profile as a result: "For all you men in your 20s, I love your hot bodies and you're gorgeous but I'm your mother's age so please don't email me."

She likes Bumble because she gets to initiate contact. "Once you connect, you decide if you want to see that person," she says. "It's quick, and most of these guys are grateful. Plus, you can filter out the creeps right away because they say something crude, which is always a turnoff in initial contact." The guys who get past all her filters have consistently turned out to be good guys. "I'll go on a date with them and if I don't connect with them emotionally, I'll still have sex with them because I love to have sex," she says. "I know that some women just want a boyfriend, but I have a complicated life."

She's speaking of her teenage daughter, Ruby. Marcelle dated Ruby's dad, who moved back to Australia around the same time she discovered that she was pregnant. "I figured I could wait until I found the right guy and then discover that we can't get pregnant," she says. "Or I could pop this baby out and raise her by myself." She was thirty-six when she popped Ruby out, and now fifteen years later, with a few fleeting relationships along the way, she's ready to, at the very least, start having sex and perhaps even find a partner.

She came close with a man she met on OkCupid. They

dated for seven months, and she thought he could be the one, but it was in the wake of losing her job and her father dying, and in those intense months, things hit a standstill. "He was good for me at that moment," she says.

Now a veteran of online dating, Marcelle has created her own rules—and fully admits that she lies about her age. "If I wrote 'fifty-three,' no one is swiping on me. But if I say 'in my forties,' I'm getting hit." She insists that prospective dates send several photos, and she looks for specificity in profiles. "A lot of men say they like, 'hunting and traveling and fishing,'" she says. "Or that they are 'looking for no drama'—very formulaic. I'm looking for someone who is a little out of the box, like 'I saw Fugazi in 1999 at the Palladium, they're cool.'"

She doesn't make quick assumptions about looks anymore, either. "At first, I was like, 'You're not hot, eh,'" she says. "But now I'm way more open." She dated a man for a while who "had a little bit of gut and silver hair," and she grew to find him "gorgeous." Another guy she dated was six feet three and shaped like "a Christmas tree" but had nice lips. "I figured I could kiss him, and we could explore from there," she says. "But I already knew I would have to be on top in bed, or else he'd smush me like a pancake."

Another rule is she will never meet any new man at his place. "First time sex is always at my place," she says. "I live in a doorman building. It's very public who I walk in with, there's a record." She prefers that for safety reasons, but it does not stop her from having fun. "My other criteria about the online dates is if I know I'm not going to go out with them again, I'm totally open to fucking them," she says.

But her biggest epiphany is that men her age are more "scared" than women about online dating and, for the most part, look more their age. "Women in their forties and fifties need to know that we look better than men our age!" she says. "We've been conditioned to put eyeliner on in the morning and brush our hair, we're in better shape, even those of us who think we have an extra thirty pounds, and we have better attitudes."

Anna, 23, on how she could date digitally after being assaulted—twice.

When Anna arrived in New York City after college, she didn't know a soul. "I met my roommates through a Craigslist apartment ad and downloaded Tinder to make friends," she says. She had also never been in a serious relationship. "I grew up in a really small rural town in Pennsylvania," she says. "And then in college, I never met anyone I actually wanted to date."

That changed when she met Jon*, one of the first men she matched with on Tinder. "He was really smart, and he seemed driven and motivated," she says. "I really liked him." They went on a few dates before she agreed to go back to his place. "I wanted to do more than kiss him," she explains. But then, as they were fooling around in his bed, she realized she was not ready to have sex. "I said, 'Stop. I don't want to do this,'" she recalls. He did not stop. "I told him he was hurting me and he just held my arms down harder. Eventually he flipped me over so I couldn't talk anymore. That was my first time having sex," she says.

After he was "finished," she quickly got dressed and went

home. There, her roommate could sense something was wrong. "She said, 'Are you okay?'" Anna recalls. "I said, 'I'm fine! I just had sex.' And she's like, 'You're not okay?' I burst into tears."

A few days later, Jon messaged her. Anna was still confused about what had happened, so she agreed to go by his place during the day, thinking she could get some clarity. "The first thing he did was hit my ass," she recalls, and she realized that she might have made a mistake. "I said, 'That's not why I'm here—I didn't have fun last time.'" But instead of apologizing, he tried to get her back in bed. She left the apartment in a panic, blocked his number, and took a break from Tinder.

A few months later, Anna told her mom, who suggested reporting it to the police. "I couldn't help but think, 'Who's going to believe me?' So much time had passed, and besides, I was drinking and on Tinder. I blamed myself for it happening." Instead of reporting it, she chalked it up to a terrible experience and tried to put it behind her.

When she finally went back on Tinder, roughly six months following the incident, she quickly met Javier*, who she says swept her off her feet. "He was totally charming." But after a few weeks of dating, he started criticizing her weight. "At restaurants, I'd order something, and he'd say, 'Careful! You'll get fat.'" Anna had an eating disorder growing up, which he knew about, so his comments weren't easily ignored. "He'd also say things like 'You're not skinny, but I still like you,' or the reverse, 'You need to eat more, you're losing weight,'" she says. "It was a constant mental game."

They had been dating for two months when he went onto her phone, without her permission, and saw that she was still on Tinder. "I was at his apartment, and he started screaming at me and then pushed me against the wall so hard that I knew there would be bruises," she says. Terrified, she ran out of the apartment, but he followed her. "I hid beneath the stairs in his building, but he found me and followed me out of the building, though thankfully, I was able to hail a taxi." On the way home, she called her parents and was shocked to learn that he had already contacted them. "He messaged my dad on Facebook, sharing details of our relationship and asking where I was. He alluded that he would be waiting for me the next day after work," Anna says. So her dad drove up from Pennsylvania to make sure she was safe. He also reached out to his chief of police, whom he knew, to see what Anna could do to protect herself. By then, she had received threatening emails and was legitimately scared, but she decided against getting a restraining order because she learned that he was leaving the country on a long business trip. Thankfully, he stopped sending her messages, and she was able to move on with her life.

However, the experience rattled her. Anna didn't think she'd ever date again. Instead, she went twice a week to a cognitive behavioral therapist to work through the trauma of both experiences. "I started to meditate," she explains. "And that really helped me see myself not as a victim, but as a survivor."

Six months later, she reconnected with a guy she'd always liked from college. "It was Memorial Day weekend, and we were both at a lake with friends," she says. "But I was too scared to even fool around." The two stayed in touch, and a few weeks later, she told him what had happened, and that experience turned into the loveliest antidote. He was finishing his

MBA, and so Anna knew that it would not be a long-term relationship, but as she explains, "He reminded me what I should be looking for."

When she was finally ready to be intimate with him, she told him, "If I say stop, I really need you to stop. I'm not teasing you." But after they talked it through, he decided that he wanted to wait—until she really felt ready. "He was awesome and made me realize how important trust is," she says. That experience taught her something profound about relationships. "A great guy is going to respect your boundaries," she says. "It's non-negotiable."

The summer ended. He went back to grad school, and she went back on Tinder. "My heart stopped the moment I matched with the same man who raped me," she says. "I immediately reported him—and left a comment about what he had done." When she matched with Javier, she reported and blocked him as well.

Anna feels way more in control now and has rules for dating on Tinder. "I always meet in a public space," she says. "If anyone ever invites me to his place, I say no. If they push back, then I don't go out with them." She is also more interested in conversations than looks as a barometer for whether she will go on a second date. That tactic helped her find a great guy on Tinder who wasn't her physical type but whom she found fascinating to talk to. "I had never had that type of chemistry with someone before," she says. "I mean you have the initial lust at first sight with some people, and then you build the other conversation in. But with him, it was the reverse."

That turned into a six-month relationship. "He taught me to

look for the whole package—someone who makes me feel good and treats me well, and someone who I can be sexual with," she says. That experience also helped her think about digital dating as a plus, an incredible turnaround after such a horrible start. "My experiences have molded me into the person I am today," she says. "Dating apps have given me the opportunity to connect with people outside of my immediate circle and have experiences with people who have changed my life—more good than bad. Whether it is Tinder, Bumble, or Hinge, I will keep using them until I find *the one*."

Rule #6

You won't get skinny by eating the same old sh*t.

An Olympic coach once told me: On days you don't feel like going to the gym, take the pressure off yourself by going with the small goal of only stretching. Lie down on a mat and start stretching, and you'll often be surprised at how once you start—and you're surrounded by others working out—you'll be motivated to hop on a treadmill or pick up a weight and then suddenly you're working out. The same goes for dating. It can feel like a massive, almost impossible effort to get yourself out there. But once you get into a rhythm, surrounded by others doing the same thing, it gets easier.

Dating is like exercise. You dread it. You don't feel like it. You would so much rather sit on the couch in cozy pj's and binge-watch *Gilmore Girls*. That is in part because we make such a big deal out of it! We have spun the concept of dating into something so high stakes that it drains any of the fun out of it. Stop the whirling in your brain for a moment and think about it strategically.

Analyze when you are at your best. Are you good at small talk and making others feel at ease? In that case a blind date may be a good occasion to shine. Or are you freaked out by the thought of talking to a stranger with the added pressure of a friend having bigged you up beyond all recognition? In which case, ask matches if they want to go bowling, try a food and wine pairing course, take a walk, or go to a concert—dates where you have something right in front of you to discuss. To take the pressure off, try a group date with other single friends and perhaps some couples. Make a game plan that puts you at your best front and center.

Dating really can suck. It can be awkward, embarrassing, humiliating. It can turn the toughest and most confident women into giddy and insecure wrecks. It can also be fun, exciting, hopeful. And if it doesn't lead to love, it can lead to you extending your social network. You can make great friends out of it whom you may be able to introduce to others. It's worth it, stresses anthropologist Helen Fisher, because what you win for putting yourself out there is, in anthropological terms, a "mating partner"—regardless of whether you want children.

"Dating is not fun," Fisher admits. "But it will lead you to life's greatest prize. You have to make time for it." And the more time you put in, the better your chances are of finding that partner. This is, she adds, the most important decision a woman will make in her lifetime. "She will have a lot of different jobs, holidays, and vacations, friends that come and go, and a lot of places where she lives," Fisher explains. "But she is really only seeking one partner, and for many women, this is the person she will share her DNA with."

Dating is the essential exercise part of your regimen. Turn it into something you do regularly. Schedule it. Kickboxing on Tuesday, JDate on Thursday, yoga Friday morning, Tinder Saturday night. Pace yourself. Try two dates a week. Mix it up. Use different apps.

Dating is the essential exercise part of your regimen. Turn it into something you do regularly.

And try to have fun.

Often when you are on a quest, it's easy to forget that the whole point of this endeavor is to bring joy to your life. We are basically hardwired to want to find a partner—and while our cultural expectations have changed dramatically over the last few decades, the cognitive wiring is still the same. "In past generations it was our parents, priests, and friends who helped us find that person to procreate with," Fisher explains. "That job has been taken over by the internet."

The great news is the world has opened up and so have our dating possibilities.

In agrarian times, a woman's mating choice was limited to those who lived within walking distance of her village or farm. That shifted in industrial times with the huge move from the countryside to heavily populated cities, where networks blossomed and strangers from all over a region would find themselves living cheek by jowl for the first time. As men went off to work in the factory, women were needed to stay at home to take care of the children and the elderly and tend to domestic life. That, too, changed as women entered the workforce. "It used to be marriage happened when you were twenty-one or twenty-two and now it's twenty-eight, twenty-nine," says Esther Perel. A Pew Research Center report found that just 42 percent of people ages twenty-five to twenty-nine were married in 2010 compared to 84 percent in the 1960s.

Author Rebecca Traister, whose book *All the Single La-dies* examines the rise in the number of financially independent and single women in the US, argues that despite their impressive educational achievements and earning power, single women between the ages of twenty-two and twenty-nine are still viewed as "pre-wives," as if they only become full-fledged and serious citizens once they partner up. In reality, she points out, those years are not only a woman's most fertile window for having children but also a fertile period for launching and growing her career, a vital part of her life. Perel says that for those women who do want to get married, that new decade before getting hitched is now spent developing and building an identity. "When you meet someone, you want someone who bestows upon you the recognition of your beautiful identity development," Perel says.

As discussed in the previous rule, that's a lot of pressure to place on one person. "The new definition of 'the one and only' has become the one who is going to make me want to stop swiping, who's going to cure me from my FOMO, quiet all my inner rumblings," says Perel. "It's the one and only in a cacophony of competing options. How do you know there isn't another one that is a better one and only?" It's a bit like a hamster on a wheel, spinning in place without going anywhere. And yet, so many women get caught in this game of keeping their options open with potential partners as they continue to search for their "one and only."

It's a state that Perel refers to as "stable ambiguity," where both parties lead each other on or send mixed messages to keep the other dangling on the line, because they are not yet ready to commit fully or give up entirely.

Sound familiar? I spent my twenties doing this to people and having it done to me, too. It wasn't especially satisfying, but I thought of it as keeping my options open. "That mentality, which is now baked into the dating world, leaves people feeling disposable," Perel says. "So you are in a relationship deeply enough that you're not alone, but not too much so that you have closed your options." Fisher adds that the idea of vast possibilities served up by the web leads to a kind of paralysis known as cognitive overload. "Women have so many options that they keep on looking for the right one," she explains. "So instead of being able to close on a potential mate, they often get pickier and pickier about who that person should be." The brain experiences decision fatigue and in the end chooses none.

Dating online is not dissimilar to shopping online, where you scroll through eighty pairs of white jeans to find the perfect pair, and when they finally arrive, you put them on and feel crushed with a disproportionate disappointment that you made the wrong choice and haunted by the ones you missed. As Perel puts it, relationships today have been commoditized. "It's not a production economy anymore, where you need six or seven children to help on the farm," she says. "It's a service economy in which you want a meaningful experience."

It's no wonder women want to play the field, consider their options, and hold out for the perfect mate. The stakes are high. But holding out too long may only keep you from discovering whom that ideal partner may be. "On the one hand, you have maximum freedom, but because you have so much freedom, you also feel that you have no control," Perel says. "You're trying to control the unpredictability of life. But life does not work like that."

Prior generations had different expectations—women knew the type of person they needed to marry, from what social class, what education, and what religion. "All those categories were mating categories and selection processes," Perel explains. "Now the world is wide open." You can date whomever you want, across race and religion and gender. It is an exciting time, and that in and of itself can feel overwhelming. Precisely why you must set your own criteria, and reassure yourself that this person is worth a second, third, or fourth date.

Fisher makes this task simple by suggesting you limit the number of partners you even consider. "Pick one potential partner, and get to know that person by going on more than one date," she says. "All the data shows, the more you get to know someone, the more you like them and the more they like you."

Besides, as mentioned in Rule #4, the idea that you will fall in love at first sight is entirely possible, but it's really not probable. In the annual "Singles in America" study that Fisher did for Match.com, she found that 59 percent of women interviewed do not believe that they will even have chemistry on the first date and that 34 percent don't believe they will have it on the second date. Yet the romantic idea of love at first sight still leads so many others to have such high expectations going into that first date that they thwart any real possibility. "Forget about the chemistry—romantic love is like a sleeping cat," Fisher says. "It can be awakened anytime, and unless you give people a chance you will be forever looking. It is as simple as that."

Stop fantasizing about the stranger who might be out there who is just right for you, and start imagining all the places someone else may be—including potentially sitting right next to you. Spend more time looking for a partner IRL than you do managing your social media. We are all prone to what I call ADDD: Attention Deficit Dating Disorder, a modern ailment triggered by the fact that you don't need to commit because you can always find another match, and so you don't focus on what can be hiding in plain sight. Of course, the failure to observe what's right in front of you isn't new. Look at the ongoing appeal and success of Jane Austen's now two-hundred-year-old novels, dripping with the exquisite tension of will they or won't they: "Will Emma get it together with Mr. Knightley? Will Elizabeth Bennet finally admit what we readers have known all along—that despite arguing with him constantly, she is head over heels for Mr. Darcy?" Those story lines are riveting because they mimic the real life in which our future loves are often staring right back at us.

"The same guy you thought was a complete drip two weeks ago might end up being one of the sexiest, funniest guys on the planet."

You have to give love a chance. We never know when it will become a possibility, so it is vital to make yourself open to every and any possibility. As for how to know who you will love, that's tricky. "We all grew up with a love map," Fisher says. "You build an unconscious list of what appeals to you as a partner. But we don't know all the things that turn us on and appeal to us. The same guy you thought was a complete drip two weeks ago might end up being one of the sexiest, funniest guys on the planet. You just have to give him a chance. There is no secret to this."

There is no other way to say this than just do it.

TO DO

Make a list of all your prospects and choose three.

For each, think of interesting ways of getting to know each other better.

Suggest a bike ride, a concert, or a museum visit. Or a walk in the park. Borrow a neighbor's dog to take the pressure off. Try a political protest or a sports lesson or a magic show.

Or you could go to a bar but maybe with live music so you don't have to talk all the time and you can observe their reaction to the band. Doing something together will give you more opportunities to really get to know each other; it will give you something in common from the get-go and allow you to check if they have a sense of humor.

If you have a pleasant time but feel there is no chemistry, *do not* promptly write that person off. It's counterintuitive, but unless you are absolutely disgusted by them, go out on another date. Love comes in all different forms. It can take time. And sometimes, it can even sneak up on you.

CASE STUDIES

Gigi*, 56, on how she never thought she'd remarry—until her longtime colleague confessed his love for her.

Gigi met her first husband when she was nineteen years old. Over the next twenty-three years, they had two children and built two successful careers as working parents. But through "the daily stresses of life," she says, "we stopped paying attention to each other."

The marriage faltered, and so they divorced. She was forty-two, their kids ten and fourteen, and Gigi never thought she would date again, let alone get married. "I felt like my marriage was a failure and just wanted to focus on my kids and my work," she says. As a senior vice president for a large entertainment corporation, she was financially independent, and she had two boys to get through school. "I didn't think that I'd meet anyone in my forties," she says. "So I decided not to try."

That did not stop men from approaching her. "One much younger guy asked if I wanted to have a drink while we were both on line at the grocery store," she says. And a married colleague started flirting with her "aggressively" at a work event. When she told him it was a bad idea, he asked why. Her answer: "Because you love your wife." Both incidents furthered her resolve to stay single.

And then one afternoon she was in a car with her boss, with whom she had worked for seven years. "We had just given a financial update to our executive team and were on our way to the airport when he turned to me and said, 'I have to tell you something,'" she says. "I thought he was going to say, 'Good job!' Instead, he said, 'You've become my best friend and I love you.'"

The admission stunned her. "We had never even flirted, let alone held hands, or kissed," she says. "I was completely caught off guard."

Plus, he was her boss, which she pointed out in the car: "I said, 'We have hundreds of people who report to us!'"

By then, the car arrived at the airport, and Gigi jumped

out and "literally ran away from him and then avoided him for the next few days," she says. There was a lot to consider—she adored working with him, but had never thought about a romantic relationship. "He's this larger-than-life, crazy, full of testosterone, independent-minded outdoorsman from Connecticut," she says. "And I'm a small Asian liberal tree-hugging city girl." Besides, he was the number one in their division; she was number two. If it did not work out, then she would be the one to leave the company, which felt like too big of a risk. She admitted all this to him, and she recalls his reply: "Isn't our life more important than our job?" He also added, "I can get another job. I have never felt this way before. I am willing to risk it."

Then he pointed out something she could not argue with—they had a head start on most couples because they already liked each other. Gigi did like him, immensely, and started to consider a relationship with him. After much analysis she decided that "the risk/reward ratio was worth it."

During their first date—a walk at a state park—she realized that she did have feelings for him as well. Three months later, they decided that it was time to tell their bosses. That's when Gigi saw a change come over him. "He went into a panic," she says. They'd been planning a work trip, and she suggested that she not accompany him, sensing his unease. When he agreed, she said, "Okay, this is over."

The two went back to being colleagues, though stopped being friends. "It was a bummer," she says. Worse, he called to let her know that he had started dating someone else. When Gigi thanked him for letting her know, he said, "Aren't you going to call me an asshole?" Gigi replied coolly, "You know exactly who and what you are. You don't need me to tell you." He admitted that was more brutal than if she had called him an asshole.

Another four months passed before he told her that he realized that he had made a terrible mistake. Gigi recalls that conversation: "He said, 'I'm such an idiot,' and I said, 'That has been established. Now, what do you want to talk about?'"

He wanted to try again, but Gigi felt that she could not trust him—and told him so. Admitting his panic—and emphasizing that he had in the interim realized that he truly loved her—made her feel confident in giving the relationship one more try, as she still had feelings for him as well.

Slowly, the two started seeing each other again. At one point, he confessed that he was worried that she was "too nice," and Gigi said, "Wouldn't that be wonderful? If our relationship was built on kindness?" He agreed. And while they had already agreed not to get married again, she accepted his proposal after one of their colleagues married his partner of many years after gay marriage was legalized. "It was such a moving story," Gigi recalls. "So moving that Shawn proposed, and I said yes."

Gina*, 32, on meeting her husband through a one-night stand in Italy.

Gina was twenty-five and running a PR firm when one of her clients invited her and her business partner Stefanie to visit their handbag factory in Florence. The two decided to go and turn the experience into a ten-day adventure. She asked an Italian friend living in New York for recommendations in Rome and he gave her his brother Mateo's number. Gina called Mateo their first night in Rome and met him and his cousin Luca at a restaurant that same evening. After dinner, they went to a super-exclusive speakeasy followed by dancing. "I started making out with Luca on the dance floor," Gina says. "And slept with him that night, thinking I'd never see him again. When in Rome!"

But the following afternoon, both she and Stefanie, who had spent an equally fun night with Mateo, called "the Italians" to see if they wanted to go out again. "It was a repeat of the night before," she says. But Mateo was headed for Milan the following day, and Luca to his beach house an hour outside of Rome.

Since it was so close, Gina texted Luca to see if he wanted to join them for dinner that third night. He did. After dinner, Gina dropped Stefanie off at the hotel and went home with Luca. "When I woke up the next morning, he was cuddling me," she says. It was hard to say goodbye. "I had this really weird feeling that he may be the man I'd marry," she says.

But that same day, she was leaving for Florence, where she and Stefanie continued having fun—and hooking up with different guys. Gina posted on Facebook a shot of one man kissing her cheek at a dance club, which prompted Luca to text: "Are you having fun hooking up with all these guys?"

It *was* fun, but nothing close to what she felt for him.

Due to a train mishap, she missed her flight back home from Rome and so texted Luca to say she wanted to have dinner and one last goodbye. They spent another night together, and this time, Gina felt really sad to leave.

That was the end of November, and he reached out soon thereafter to say he was coming to New York that December. "That weekend, we fell in love," Gina says.

She next flew to see him in January, and then he came to New York in February. In between they Skyped (this was before FaceTime) and talked at least once—sometimes multiple times—a day. "We got to really know each other," she says.

Then he called in April to say that he had an opportunity to invest in a restaurant that was opening in New York City and would handle his visa. That meant moving to Manhattan, but not until September, which felt like an eternity. For months, there was no news on the visa, which made Gina nervous that it might not work out. "I started to explore what else was out there," she says. "I just did not feel like it was real."

She still had feelings for him but wanted to be realistic. That summer, she did go spend three weeks with him in Italy. And while it was "bliss"—daily sex in a picturesque village—it also did not feel real. So they started talking in practical terms about his coming to New York. With no visa news, marriage felt like their only option to be together. "That was not the way I wanted to get married," she says. In the end, the visa came through, and Luca moved to New York in late September, after the two had been together for almost a year. He got his own apartment to start, though spent every night with Gina. A year later, they moved in together. Roughly six months later they woke up one morning and looked at each other and said, "Let's get married."

Rule #7

Stop with the comfort foods. It's okay to be a little hungry.

Once you start to live with a healthy love diet, you want to be able to resist temptation in your weaker moments, so you need to create an environment where you can't fall back into those old bad habits.

If you love donuts, but you know that they will inspire remorse after the initial thrill of eating them, keep them out of the kitchen. Don't drive by the donut shop on your commute, and delete the number of the bakery that delivers on your speed dial.

The donut is a metaphor for your ex—warm, sweet, familiar, and loaded with trans fats that clog the arteries and eventually lead to a blockage of the heart. If you're in a friends-with-benefits situation, then ask yourself who is really getting the benefit. The ex and the FWB are the equivalent of relationship snacks. They stave off hunger in the short term and give you a sugar high for a brief and comforting moment; then comes the inevitable blood sugar low, followed by the crash and burn. Next time around, your hunger grows more acute.

You made the list of tempting treats in Rule #3. Now go back, revisit the list, and recheck your cupboards. Are you sneaking stuff back in there and pretending it doesn't really count?

If your ex continues to ping you for a Tuesday-night booty

call or your study mate wants to have sex before and/or after your note-review session, don't let it become a habit unless you're sure that's what is in your long-term interest. Instead, pause and ask yourself, "What am I getting out of this? How will it make me feel during? Afterward?"

We all have cravings, physical and emotional, and sprinkles and glaze may address them in the moment, but we're planning for the long run. And don't kid yourself by saying you will eat only half the donut—or just this one time it doesn't matter, you won't do it again.

The donut is a metaphor for your ex—warm, sweet, familiar, and loaded with trans fats that clog the arteries and eventually lead to a blockage of the heart.

To stay on the analogy for just one more moment, it's easier to skip the cream-filled donut entirely because a single bite is engineered by teams of food chemists to make you crave the whole thing and then lunge for another one. This is not an equal struggle! The mouthfeel of warm, smooth paste is fighting not only your own willpower but also all the brain chemistry triggered by the sugar and corn syrup and the fats so carefully engineered to produce a reaction.

Similarly, when you have sex with people, it unleashes all the feels. That's what sex does. Especially with someone whose buttons you know how to press and who can press yours back. It's too hard! Easier to stay away, block their number, find new hangouts where you know you won't run into them.

Give yourself a chance by giving yourself a clean break. You don't need to hang on to an ex. Sometimes you need to be brave. It requires quitting, cold turkey.

The same is true for the sexless trysts, the crushes, those unrequited obsessions that consume our emotional energy and imagination yet leave us depleted. Count up all the hours you have wasted hunched over your laptop obsessing over your exes' feeds—it starts as a quick, casual check, and then the pattern of a solitary night in is set: nonstop checking fueled by insecurity and jealousy. Social stalking gets you nowhere and leaves you angry and empty. Think of the energy and time you have wasted, the books unread, the people unmet, the hobbies untried, the gym unused, as you myopically slip down the solitary Instagram rabbit hole logging whom they've tagged, not to mention the countless hours cross-referencing who has tagged them back.

In the same way you pick idly at chips promising this is literally your last one, you may be in a relationship that you know isn't going anywhere, but you're hungry for love, and it feels less frightening than nothing. You're busy, and it's easy; it keeps the engine running. Maybe he's married, so it's not a possibility in the long run. It's just a distraction.

So before you have any more hookups or go on any more dates, do another inventory. Make a list of all the people in your life and rate them in terms of energy in, energy out. Is there anyone in your life right now who is blocking your love quest? It might be an ex or a neighbor or an old college friend. It may also be your best friend, your parents, or your brother.

This may be the best girlfriend who gets twitchy and hypercritical when you start dating someone. It could also be the overbearing boss who demands you stay late or work weekends during the times you should be out exploring other arenas of your life. Or it could also be the demanding or dysfunctional family members who view your finding love as a direct threat to their well-being, or who want to persuade you to settle for someone who makes their own lives more comfortable.

Write down the name of each person whose opinion is getting in the way of your love goal. And then make a note to stop sharing the details of any new relationship with them until you are sure the relationship is solid enough to withstand their feedback. Oversharing with your friends is so damned easy, and once a secret is out there, you can't wrestle it back. This goes for conversations as well as putting intimate details online.

There's a moment in the first season of Issa Rae's brilliantly observed *Insecure*, a dramedy on HBO about the life of a single twenty-something woman living in Los Angeles, where Molly, a successful lawyer and Issa's best friend, falls for a gorgeous, kind guy who works as the manager of a car rental company. By the time we meet her, Molly has been on many online dates with assholes, but this guy is different. He treats her with respect. He makes her feel great about herself. They have real attraction—both physical and emotional. So when she asks him to share his most intimate—and embarrassing—moment ever one night after splitting a bottle of wine, he admits to a drunk dalliance with a male friend when he was in college. She freaks out. And in the next scene we see her regaling her gaggle of opinionated, gossipy friends with his secret confession.

The viewer knows instantly Molly has ruined any chance of making that relationship work because to move forward with the guy will be against all the opinions of her judgy friends. Women do this all the time. We give our girl squads—fun though they are—too much information and power. We invite them to play therapist, but without the training or objective distance a therapist has. We fail to ask ourselves the hard questions—why are we with him? Does he make us happy? Who cares if he once drunkenly fumbled with a guy if it was ten years ago and he makes you feel great? Too late. Your friends have already condemned him. But they don't always know what's good for you.

Private oversharing with girlfriends is one barrier to firming up a new relationship—another barrier is social media, which creates a bogus urgency to publicly overshare. Most of us do it. We overshare the most banal details—the waffles we ate for breakfast, the new Olivia von Halle pajamas we wore to bed, how much we lifted at the gym (guilty!), whom we made out with while drunk at the club.

On one level, it's supposed to make people feel connected to you, included in the journey of your day. On another level, it's a profound search for validation. And when you spend too much time on it, it turns you from a participant in your own life into a performer in search of an audience.

How many friends do you have on Facebook? On Instagram? And LinkedIn? "You're not the same person on Facebook that you are on Instagram that you are on LinkedIn," cyberpsychologist Mary Aiken adds. "So you're presenting all of these selves and it's a lot of work and maintenance. It's exhausting." It also leaves little time or energy to nurture the work of a real relationship. Spend time on developing actual friendships. "Dunbar's number dictates that the maximum number of relationships we can cope with as a species is 150," Aiken says. "After that, we begin to suffer from social exhaustion."

Like a fragile sapling, a new relationship needs space and light to grow strong. Relationships are built on mutual loyalty and respect. And that means keeping the intimate details, the moments of deep sharing and trust, between yourselves.

So that means identifying all the people in your life who want you to be *truly* happy—and still, sharing any future partner with those people only when you are ready to.

Don't turn dates into anecdotes to entertain your friends unless you are sure you're not going to see the guy again. Keep something on the plate for later.

CASE STUDIES

Louise*, 38, on knowing when it was truly over.

Louise, a marketer, and Martin, an engineer, went out for four years in their early twenties. "I had known Martin a little at high school, where he was two years older than me and a total catch, and we ran into each other randomly after college and started dating, and I was over the moon. I was completely in love with him for the first two years," Louise says. "He was my first real boyfriend, and we had a fantastic time together. We had this amazing friend group, we traveled together. He was my lover, my best friend, I loved his family, our mothers got along, we talked about our future together. And then after about two years I slowly began to realize I was more ambitious than he was. That I wanted to explore more things, better work, new friends, and he was happier just spending time with me at home or with his old familiar friends. I had a recurring dream that I was a sailboat and I was trying to sail faster, and he was an anchor, slowing me down.

"He turned from this guy who'd always seemed ahead of me into someone who was now behind me.

"It took me a year to admit this to myself. Partly because

everyone assumed we were happy and because admitting we weren't, even to my mom, would have been disloyal to him. And among my friends, I was the lucky one, the one who had found someone already, and I didn't want to give up that status. And it was partly because I couldn't bear the idea of telling him because I knew that for him, this was it. He'd found me and he wanted to keep me, and I didn't want to disappoint him. I finally told him I needed a break, and he was devastated, but then to try and prevent too much pain we kept saying we would stay best friends, and it was just like we couldn't actually split up.

"It was partly my fault. I knew I didn't want to marry him, but if I was staring at an empty Saturday night, I would call him. And he would get mad and start telling me I wasn't being fair. Which I wasn't. But then we would go out for dinner and I would think, 'Well, maybe I'm wrong, and we should stay together.' And we'd have great sex and go out for Sunday brunch, and it would feel so reassuring instead of facing a very uncertain, lonely weekend. I hadn't found anyone new yet, and I was scared that maybe I wouldn't, and so I was in this emotional no-man's-land. I was trying to persuade myself to leave and get on with a new life, but at the same time I was too nervous to make the leap I needed to, so I would come running back to him."

This EWB (ex with benefits) situation lasted for about a year. "It was harder to find new people than I expected—and then one night we sat down for a drink, and he told me he had accepted a job in Brazil, which meant we would have to split properly because he couldn't go on like this. Even though I knew I was eventually going to meet someone else, I still remember the internal panic when he told me. But I also knew he was right and I couldn't stand in his way. He would move abroad a month later, and we had sex for the last time. It was so familiar and good and caring and so sad.

"I changed jobs, I bought an apartment, my life started growing in the way I had hoped it would. And he met someone in Brazil. I dated a few men, and then about five years after Martin moved, I met the man I would marry and have two kids with.

"After ten or so years, Martin got a huge job and reached out unexpectedly when he moved back to the US and suggested the four of us go out for dinner together. I thought about it, but I never replied. I thought I might be toxic for him, and I didn't need to see him with someone else. We were done."

Amy, 35, on turning a devastating breakup into a business.

Amy was twenty-seven, living in Vancouver and the director of marketing for a luxury hotel company, when she met the man she thought she would marry. They were introduced at a party by a mutual friend and began flirting on Twitter for a few months before they finally started dating.

A year and a half later, the two moved in together and started talking seriously about their future. He was an entrepreneur who had started a company, and while she had a blossoming career in marketing and PR, she envisioned staying home once they had children. "I wanted to be flexible around his work schedule," she says. Her mother stayed at home while her father worked, and Amy based her future life on that model. "I knew we would eventually have kids, so I stopped excelling in my corporate job," she says. "Everything was about him." She even started "practicing" how to be the perfect wife. "I entered that relationship not even being able to boil eggs and learned how to have dinners waiting," she says. "I even packed him lunches."

She was twenty-nine when she lost her job due to major budget cuts and wound up treating her boyfriend to a European trip with her savings. It was meant to be a romantic getaway—but instead led to their breakup: "I discovered infidelity and was completely heartbroken," she explains. It caught her by total surprise—suddenly she was without a house, job, or partner. "I spiraled into a depression, started having panic attacks, stopped eating, and lost about twenty pounds in less than a month," she says. "I had thoughts of suicide."

Stunned by the physical and emotional impact of her breakup, she sought answers in therapy, Reiki, yoga retreats—even psychics. While it eased the immediate pain, she was still brimming with anger and resentment, which made dating new people hard.

Then one day, as she was retelling the story to a friend, he asked if she had any good memories of her ex. "In my anger, I forgot everything that was amazing about our relationship," she says. She also neglected to consider the part she played in its dissolution. "That same day, I wrote my ex a letter taking accountability in what happened," she says. "That set me free."

Amy moved to New York for work, back in marketing, and began penning an online relationship column, too. "So many women would write to me in pain over their breakups," she explains. "I began to see a pattern." She began researching what happens physiologically when people break up and learned that love literally acts like a drug—and a breakup can trigger "an intense physical withdrawal," she says. "Plus, there's so much shame behind it."

Amy realized that there was a need for a place for women to share their thoughts and feelings about bad breakups, without being judged, so she organized a weekend "breakup boot camp" retreat where she invited a psychologist to talk about what happens in one's nervous system following a breakup, plus what exercises to do when feeling triggered. "Like how to avoid Facebook stalking because you want that next dopamine rush," Amy explains. "You really are physically in pain after a breakup—but it gets better with time."

Amy is proof of this. Due to huge demand, she is currently planning the second workshop and has plans to make this her full-time work. And while she has not yet found her life partner, she feels confident that she will. "I really believe that patience is key," she says. "This time, I want to make the right choice."

PART THREE

SHORTCOMINGS AND LOVE TRAPS.

Rule #8

Alcohol is not a food group.
Respect your limits.

Question: According to FBI and police officers, what is the most common date-rape drug in the US?

Answer: Alcohol.

No one loves a glass or two of champagne more than I do (okay, sometimes three, but that's a special occasion), but the truth is that drinking is good fun until it's not—and finding that in-between spot is different for everyone. Some women get a headache after the second margarita and stop; others can drink cheerfully until they have lost count and consciousness. And while that decision is 100 percent yours to make, we live in a world where alcohol and sexual assault are intertwined. As mentioned earlier in Rule #4, the National Institute on Alcohol Abuse and Alcoholism (NIAAA) states that one half of all reported sexual assaults involve alcohol consumption by the victim, perpetrator, or both.

Dating and meeting people is hard. You're not in control. You're nervous about being judged and criticized by others, and you're being self-critical. Alcohol protects you. Or at least that's what the first couple of glasses feel like. It's social armor. And then there is peer pressure, which, contrary to what everyone thinks, doesn't stop at the end of your teenage years. It continues throughout your adult life. Everyone else drinks! Why wouldn't you join in on the fun?

The problem is it's easy to lose control. Of course, no one ever deserves to be taken advantage of—whether merely buzzed or blackout drunk. But women process alcohol differently than men. "If a man and a woman have anywhere near equal the number of drinks, the woman will get more affected and more quickly," says Sharon C. Wilsnack, PhD, an expert on the effects of alcohol on women. Men are typically bigger than women, but beyond sheer size, men have more muscle, and women more fat, which means alcohol runs through men whereas it stays in women longer. Men also have a more active stomach enzyme called alcohol dehydrogenase, which breaks alcohol down. "It's more concentrated when it enters the woman's bloodstream because it hasn't been metabolized to the same extent it has been for men," Wilsnack adds.

This higher concentration of alcohol leads to more than slurring words or stumbling around. "High alcohol consumption impairs judgment: so perceptually you're not picking up the same cues of danger," Wilsnack explains. "And physically you are often not coordinated enough to react." Which is why drunk women are at a greater risk for accidents, STDs, unplanned pregnancy, and sexual assault. And yet, alcohol is also a powerful social lubricant—and for most of us, intertwined in dating and romance. "In the early stages with light or moderate drinking, alcohol is an aphrodisiac," Wilsnack says. "It does make us more relaxed. That's the hook for most women."

It's fine to have two glasses of wine over the course of an evening—but no matter how strong you think you are, four gets most women to a point where it may be hard to speak clearly or walk straight. (In fact, NIAAA defines binge drinking for women as four drinks in two hours or less versus five drinks in that same time period for men.) The key is knowing when to stop—because that buzz we seek on a Saturday evening can quickly turn into something far less fun.

No one has written more thoughtfully on this than Sarah Hepola. Her memoir *Blackout: Remembering the Things I Drank to Forget* is a wonderful explanation of how someone can so easily get in over her head using alcohol to help find love. Alcohol became her consent and allowed her to have sex with people she might not have slept with while sober, to avoid feeling lonely. Instead of finding love, she'd wake up massively hungover, tangled up in someone else's sheets, wondering, "What the hell am I doing here?"

Does this sound remotely familiar? One of the most disturbing trends I've witnessed over the last decade is the rise of binge drinking. In 2015, NIAAA reported that of the 60 percent of all college students who drank alcohol, two out of three engaged in binge drinking. And while binge drinking has only risen 4.9 percent among men, the practice has risen 17.5 percent among women between 2005 and 2012.

When I was at college in the eighties, admittedly in the UK, where drinking patterns are different, I don't remember a single person getting blackout drunk. It wasn't a phrase you ever heard. Granted, it has become a huge issue there as well as here, despite the fact that the drinking age there is eighteen. In the US, you can't buy alcohol until age twenty-one—one of the many reasons young women pregame. Pregaming, in turn, is a factor in the rise of binge drinking, which increases a young woman's risk of sexual assault and physical injury as well as emergency room visits. According to a *Washington Post* analysis, in 2013 more than one million women wound up in emergency rooms as a result of heavy drinking. Shockingly, the rate of alcohol-related deaths for white women ages thirty-five to fifty-four has more than doubled since 1999.

Part of the reason heavy drinking has become such an epidemic among young women is due, I think, to the pressure women feel to be actively sexual with people they barely know, while also yearning for real intimacy and connection. My good friend, the former president of an all-women's college, agrees. She told me that her students admitted that they got drunk on purpose before going out on the weekends. When she asked them why, she told me, the general response was "Because I am going to take my clothes off in front of a perfect stranger, and I don't want to deal with that." She added, "It's deeply connected to the type of sex they anticipate happening. In my generation you used alcohol because it got you tipsy and flirty but not blackout drunk, because you wanted to remember it. They don't want to remember it; they want it to happen and be over with."

When I was editor at *Cosmopolitan*, we published several pieces highlighting both binge and sexual drinking on college campuses where cabdrivers would buy and supply a steady stream of beer and cheap vodka to those not yet twenty-one. As we heard from more and more women about their empty and often scary experiences, it struck me that binge drinking was essentially women self-medicating to handle the immediate sexual demands from men—with no pressure for either party to follow up the next day—and the disappointment that it all should be more fun than it is.

As it turns out, alcohol works as both a social *and* sexual lubricant. "It really does have physiological effects that can be interpreted as sexual enhancement in small to moderate amounts," Wilsnack says. "But in large amounts, you just get blotto and you don't feel sexual or anything else. That's when the assault usually comes in."

A study published in *Archives of Sexual Behavior* in 2016 found that while women felt more attractive while buzzed, drinking also could lead them to pick "atypical" partners. That same study found that alcohol made women feel more "adventurous" when it came to sex, and led men to "aggressive" behavior. One man interviewed said, "It feels like you get a lot more primal . . . a feeling of needing something and you're going to do whatever."

I know it's not politically correct to say watch your drinking, but watch your drinking. We are not letting men off the hook here—of course they shouldn't be taking advantage of you, or anyone—but if you staggered unexpectedly into the road drunk and into the path of an oncoming car, you would be more likely to get run over and the driver would be more likely to get off even if he's been driving recklessly. So respect your limits and avoid putting yourself in harm's way. It's never empowering for either sex to get blackout drunk. David Jernigan, PhD, an associate professor at the Johns Hopkins Bloomberg School of Public Health and director of their Center on Alcohol Marketing and Youth, compares it to walking on an icy sidewalk: "The icy sidewalk isn't the reason you're going to fall, but if there's ice on the sidewalk, the fall is much more likely," he says. "Alcohol isn't the reason for the sexual assault, but if alcohol is involved, sexual assault is much more likely."

Let me repeat: This is not a green light for men to get away with bad behavior. But ultimately you need to look out for yourself.

The *Journal of Studies on Alcohol and Drugs* published a study that found that 82 percent of college students who had unwanted sex were under the influence of alcohol, while another study done by the State Council of Higher Education for Virginia found that 47 percent of college-aged women who were raped in that state believe they were unable to effectively resist as a result of their own alcohol use.

Know the facts before you go out and get hammered.

This also speaks directly to the blurry lines around consent. During the summer of 2016, I was sitting in the back of a cab in midtown Manhattan when an advertisement popped on the screen embedded in the back of the driver's seat. The ad had Zoe Saldana saying, "There's one thing I give to everyone I have sex with." She was followed by cameos of a half-dozen solemn celebrities, including Nina Dobrev and Josh Hutcherson, repeating the word *consent*.

All I could think was, "How have we gotten to the point where consent needs its own advertising campaign?"

Perhaps it is because the dominant public discourse today around sex is not about love or the search to find a lasting partner; it's about rape and assault. Sex is not this fantastic, blissful experience that launched a million movies and novels and poems and paintings. It is violent and scary. It's drunken and blurry. Whether it's Brock Turner, the Stanford swimmer who was convicted of raping a twenty-three-year-old unconscious woman after a frat party, or *Law & Order: Special Victims Unit*, where each episode is based on another hideous sexual assault.

And then there are the many ways alcohol is portrayed on the big and small screens, which I can't help but think influence our attitude about its role in our lives. Take that scene in *Trainwreck* where Amy Schumer is on the phone with her friend Nikki the morning after she's spent the night at the apartment of the doctor, played by Bill Hader, whom she has a crush on.

AMY: I slept at the doctor's place last night.

NIKKI: You never spend the night. What were you, blackout drunk?

AMY: No, I had like two drinks . . . Three, max . . . Four, now that I'm tallying.

NIKKI: Cause you're on antibiotics or something?

AMY: Oh my god, he's calling me.

NIKKI: Why would he call? You guys just had sex.

AMY: [answers phone] This is Amy. I think you butt dialed me.

AARON: No, I dialed you with my fingers.

AMY: [to Nikki] He called me on purpose.

NIKKI: Hang up! He's obviously like sick or something.

AARON: I was calling to say I had a really good time last night and was wondering if you wanted to, um, hang out again.

NIKKI: I'm going to call the police.

There is little to no indication in contemporary popular culture that sex can be fabulous, life enhancing, or fun—or that it can lead to love. The only fun women in movies have with sex seems to be laughing about it afterward with their friends, as they swap their own Amy Schumer–esque experiences. Alcohol is often involved in these scenes, and everyone is laughing whenever alcohol shows up on the screen. Think *Bridesmaids* when Kristen Wiig gets really drunk on the plane and we all find it hilarious. Or the menopausal Rita Wilson and Ali Wentworth talking about their vaginas closing up over chardonnay in *It's Complicated*.

To that point, white wine has become the signifier of female bonding—both on the screen and in real life. When was the last time you went to an event where alcohol was not served? Whether a book club (sauvignon blanc), baby showers (prosecco), girls' nights out (margaritas or manhattans, depending on the season). And then all the weddings, birthdays, or any other excuse to pop the champagne and officially get giddy. A popular hashtag in summer 2017 was #RoseAllDay. Drinking is a way of showing that you like to have a good time and go with the flow. And getting drunk is a way to reinforce to the world that you are having a good time—even when you are not.

In another scene in *Trainwreck*, Amy Schumer's character chugs Bandit boxed wine, also known as "binge in a box." Trinchero Family Estates, which produces the wine, saw an opportunity and promoted the scene on social media, which prompted young women to share photos of themselves chugging Bandit. This is murky territory, as it went against the rules around advertising alcohol—the Distilled Spirits Council of the United States, one of the largest US trade groups, rejects ads that "in any way suggest that intoxication is socially acceptable conduct." The beer and wine trade groups have similar rules, but they are easily broken and, according to Jernigan, not readily enforced. Women are being marketed to specifically; think Skinnygirl vodka (PS, it's about the same amount of calories as regs vodka!) or Mommy's Time Out, a wine that comes in white, red, and pink. Urban Outfitters sells a wineglass big enough to hold an entire bottle and etched with the line "Drink until your dreams come true."

These messages start early and have an impact. "Eighty-five percent of all drinkers in the US started drinking before they were twenty-one," Jernigan says. For decades, beer was the favored choice among teens, but that has shifted recently. "Bud Light is still number one, but alcopops—Smirnoff Ice and Mike's Hard Lemonade—are numbers two and three," Jernigan says. "They are sweet, fizzy, and brightly colored, and have been marketed by the industry for new drinkers who don't like the taste of alcohol." The natural segue is from the Skinnygirl vodka of your calorie-counting twenties to the "mommy juice" wine of your thirties. *"Sex and the City* was a real turning point—those women were rarely without a cocktail," Jernigan says. "We're seeing more drinking acts per hour in entertainment today than we've ever seen before."

Binge drinking, as already noted, is on the rise, most noticeably among women. Between 2005 and 2012, the rates for women grew by 17.5 percent, and for men by 4.9 percent, Jernigan says. This is not an area where we want gender equity. The irony, and the slippery slope, is that while women get drunk more easily than men, they are also more prone to suffer brain atrophy, heart disease, and liver damage from heavy alcohol consumption, according to the Centers for Disease Control and Prevention. "Women who drink also have an increased risk of breast cancer," Jernigan says. "Women are at higher risk for liver, brain, and heart damage than men who drink comparable amounts. There is also a link between alcohol, which is a depressant, and depression and anxiety. For women who have either condition, drinking will only exacerbate it. "Adolescent girls are twice as likely to be depressed as adolescent boys," Jernigan adds. "And there's a stronger relationship between depression and substance abuse in girls than boys."

Yet on TV and in the movies, women are usually portrayed as funny when they get drunk—never pathetic or vomiting, or even if they are, still somehow funny.

In real life, what starts funny can quickly turn to pathetic and scary. Lena Dunham is one of the few writer-directors who tackles alcohol with realism. In the final season of HBO's *Girls*, Hannah, Dunham's character, gets so drunk that she ends up sleeping with the surf instructor she is supposed to be writing about for a magazine article, and then vomits over the side of his bed. Whereas Amy Schumer makes drinking humorous and, in theory, empowering for girls, Dunham shows the flip side of it, the lonely sadness of the drunk girl. Her drinking allows her to fast-forward intimacy with the guy played by Riz Ahmed—they spend that weekend together, and she starts to fall for him, when it is clear all he wants is vigorous and easily available sex. When she mentions that she might rent a place in this seaside town to work on her writing, the subtext is clear—she wants to spend more time with him. So when he says, "Oh, that's great! You can meet my girlfriend," her disappointment is painful to watch.

Alcohol hurtles you past the awkward stages of getting to know someone, which you think you don't need, when you often do. These stages actually protect you and allow you off-ramps along the way that you don't necessarily notice when you're tipsy. Alcohol blocks out potential red flags.

Sarah Hepola relates. "At the age of thirty-two, I was the one who was making the jokes about having sex with some guy and not really knowing his name and saying that it was not that big of a deal," she says. "I don't think I was out there just wanting to have these sexual experiences. I think I wanted each of those things to turn into more. I wanted a deep and sustaining love with someone else. But in absence of that it was like, this will do."

And while looking for that, she put herself in danger. Though she doesn't remember the details, she does recall the bruises the next morning. "I spent a lot of time in my twenties beating myself up for that, thinking, 'Why was I so stupid?'" she says. "But actually, if you think about how alcohol works, it is not that you were stupid. It is that you were drunk."

Sober since thirty-five, Hepola, now forty-two, is currently single and can articulate better than anyone I have met the tricky relationship between love and alcohol. The problem with dating, she says, is that there is this idea that women are supposed to just organically find love. "It's supposed to just happen," she says. Since it does not "just happen," alcohol provides the usually false hope that it can help get you there. "Alcohol gave me this sense of confidence I badly wanted and a sense of comfort in my body and a sense of exhibitionism that can really be rewarded," Hepola explains. "Alcohol gave me access to what I wanted—but then what I wanted changed, because I was drunk."

Drinking speeds up familiarity, which is encouraged by our culture, where everything else is so sped up. You can join Tinder and go out on a date with someone within two hours. Most likely, you meet at a bar, have a drink, which gets rid of your inhibitions, so you drink to become the witty smart-ass personality you created online. You also want any reservations you feel bubbling up to recede, because if you say no to sex on the first date, then you're being prim. So you drink away this worry, too, not realizing that you are also drinking away your ability to hear alarm bells if this is not a good situation. Hepola had more one-night stands than she can remember. "It makes you feel good in the moment," she says. "And terrible afterward." Just like fast food.

Her sobriety finally helped her understand real attraction. "One of the biggest revelations for me was, 'Oh, this is what chemistry feels like,'" she says. "You can tell when someone was the right person for you to be with. Whereas before, I would drink myself to that place." For college students in particular, alcohol has become central to the sexual experience. "It is seen as a lead-up to hookups, and necessary foreplay," Hepola says.

The culture around pregaming, drinking on an empty stomach, inserting vodka-soaked tampons, and lining up shots, are all risk factors in blacking out. We teach women to cover their drinks so no one slips something in them, but not that the drinks they buy themselves are dangerous. In addition to the host of risks already mentioned, women are more prone than men to blackouts, which is when your brain shuts down because you drank so much. "Drinking has been tied up with these messages of empowerment," Hepola says. "But drinking to oblivion means you have lost your power, as well as balance and judgment. And then your memory starts to shut down. That is the message that has to be disrupted. For men and women."

Especially since we can barely get through the week without reading about some horror story that involves alcohol and sexual assault. It's hard to imagine a more deeply disturbing story than that of Marina Lonina, who livestreamed her friend being raped on Periscope. In her court testimony, Lonina at first claimed that she grabbed her phone to film it as evidence but then got swept up by how many "likes" the video got. Both she and the victim had been drinking—newspaper reports claim the victim was "heavily intoxicated." The rapist, someone both women knew, had been drinking with them. He was sentenced to nine years in prison, Lonina to nine months in jail for "obstructing justice," and her friend has become the one in four women who get sexually assaulted each year.

Would the evening have been different without alcohol?

So, how have we gotten here, and how do we reassure people that a robust sex life can be achieved without copious amounts of alcohol? And remind folks that an orgasm will help you sleep far more effectively than any Ambien? And above all, that a good sex life is free? We need a better understanding of the benefits of a good sex life and a road map to finding someone you trust enough to have it with. We know that in an era of Tinder, Hinge, and Happn, finding someone to have sex with is not the issue. As one *Cosmo* staffer assured me when she dropped off some copy one afternoon, no one has any issues getting laid anymore. Finding commitment is the challenge, and having good sex, sexy sex, sex that doesn't leave you in need of a hot shower wondering whether to call the campus police, is hard.

So, get out your journal and start taking note of your own alcohol intake:

> **How often do you drink?**

> **How many glasses per night? Per week?**

> **Why do you drink?**

> **Has anything bad happened to you as a result of your drinking?**

> **If you've ever been assaulted, was alcohol involved?**

Own up to your drinking. And then do a friend-to-alcohol audit:

> **Who drinks the most of all your friends? Do you drink more around that person? Is there anything about their behavior that makes you feel uncomfortable?**

> **Who drinks the least?**

I'm not saying don't drink. I get that it's confidence in a glass—that's why we all love it. It's fun, it gives you a great buzz, everybody else is doing it, but know your limits.

What is your tipping point? Is it three glasses of wine over the course of a long dinner? Or two martinis during a cocktail hour? Know the signs before you get up and realize you can't walk straight or think straight, either. If you can't trust that your friends will cut you off when you have had enough, then you have to be your own advocate.

ACTION PLAN

It's simple: Drink and have fun but don't get drunk. More than three glasses of wine in an hour is considered binge drinking.

Or here's a radical thought, try SWS. Sex while sober. A study in the *Journal of Sex Research* found that since alcohol is a depressant, during the sexual act it can actually dampen sexual response in both women and men. In short, your best orgasms happen while sober.

Rule #9

Hookups are like french fries.

It's time to break out the sex scale. It can tell you what is out of whack.

In many ways, your sex life is one of the most important parts of this diet. That's because of all the emotions tangled up with it. It's disappointing to have bad sex with a good person, confusing to have good sex with a bad person, and depressing to have bad sex with a bad person. Hookups are like french fries: delicious in the moment, but they often lead to remorse.

The goal is to have good sex with a good person. So let's start by analyzing your own sexual history. Break out the journal and answer the following:

When was the last time you had good sex?

Who was it with?

What were the circumstances?

Did you have an orgasm? Did you see that person again?

Now, think back over your history of sexual encounters.

Have you ever had casual sex? How often do you have it?

Do you have sex to figure out if you like someone?

Do you have sex drunk? How often?

Do you regularly have orgasms?

I realize that some of these questions are intense to start off, but for all the white noise and cultural analysis around hooking up, an illusion persists that sex is casual. That it is no big deal. That everyone is having it, all the time. None of that is true. In fact, a 2016 study in the journal *Archives of Sexual Behavior* found that millennials are more than twice as likely to be sexually inactive as Gen Xers.

Gail Dines, a Wheelock College academic who tours college campuses to talk about the impact of porn on relationships, says that her female students talk about the "emotional hangover" they have after hookup sex.

One reason stated in the study was a focus on education and career. Another is all the confusion and anxiety around consent. Gail Dines, a Wheelock College academic who tours college campuses to talk about the impact of porn on relationships, says that her female students talk about the "emotional hangover" they have after hookup sex. "The students I talk to say they rarely get orgasms through hookups," Dines says. "So I ask, why even do it?"

Sex is so confusing for so many women, especially those who are in their sexual prime in their midtwenties, which as we have discussed earlier, coincides with the huge emotional effort of embarking on a career. "The pressure on women to be successful complicates things," says New York–based sex therapist Logan Levkoff. "It makes it seem like having an emotionally intimate relationship takes so much time that we need to postpone it for the future—when we get our lives together."

Relationships do take energy and time—but so does hooking up. "Casual sex isn't always casual," Helen Fisher says. "It can trigger a host of powerful feelings, as any genital stimulation can cause an increase in dopamine and any tactile stimulation can cause an increase in oxytocin." Both lead to romance and emotional attachment, and Fisher wonders if people engage in "hooking up" to unconsciously trigger those feelings.

As with everything else in this diet, I am advocating acute consciousness when it comes to sex. Mindless munching leads to empty calories—and unwanted pounds. Likewise—to paraphrase Gwyneth Paltrow slightly—unconscious coupling can also weigh us down and get in the way of our love goals. So before you climb into bed with someone new, pause for a moment and reflect honestly about what your expectations are.

If you want to hook up just for the pure fun of it and it feels liberating and you don't need it to go any further, then great.

If your goal is to see him again, and you think that that's his goal, too, and having sex marks the beginning of a mutual relationship, that is exciting as well.

But if your hope is that sex may lead to the start of something but you're not sure, then you should allow for the possibility that you may be disappointed if they don't text you until the next time they want a hookup.

Before you climb into bed with someone new, pause for a moment and reflect honestly about what your expectations are.

Whatever your goals, once sex is involved, fasten your seat belt and prepare for emotional turbulence.

I don't want to sound as if all hookups are bad. Of course they're not, and they can lead to something bigger. "We often think of casual sex as being in the domain of what men want and women go along with it," says Justin Garcia, the author of a study published in "Sexual Hookup Culture: A Review." His research has found that 63 percent of men who engage in hooking up say they would prefer a romantic relationship. "Many men think that they're supposed to hook up," he says. "But they really want a relationship as well." He also found, in a collaborative study he did with Helen Fisher for Match.com, that one in three people surveyed had a sexual hookup that turned into a romantic relationship. And to further debunk the myth that all men just want sex with no strings attached, Garcia cites another study he did on cuddling: "Over 50 percent of both men and women want to spend the night and cuddle after a hookup," he says. "These are not the cultural rules of no-strings-attached casual sex."

The data, he says, leads him to believe that some people are looking for—and actually getting—aspects of intimacy in their sexual hookups. And he argues that a hookup, or having sex early in a relationship, is a good way to gauge your sexual attraction to a person, which can actually change during sex.

"You're invoking all senses when you have sex with someone," says Ian Kerner, PhD, psychotherapist, sex counselor, and author of the wonderfully titled *She Comes First* and its companion *He Comes Next*. "So you're getting all this biological information about this person and your own body is having a biological response. Do you like the smell? The taste? Do you like how it feels? Do you feel safe? Did you have an orgasm?" All this important information is explored during a sexual encounter. "You learn a whole lot about the person," agrees Fisher. "You know instantly if it is going to go forward—if they don't call you the next morning, you know it won't work. And it's better than spending eight months talking to the person and then having sex and then realizing it is not going to work."

By "work," Fisher means, is this person someone you can explore your sexuality with? Whom you can imagine being truly intimate with? That is what good sex is.

Intimacy is the key ingredient. It is the organic roasted sweet potato that takes time to find and prepare. "People crave intimacy," says Kerner, who also uses the phrase "Pleasure is the measure" to define any healthy relationship. "Without it, you end up with a sex life that becomes somewhat dehydrated or desiccated," he explains. "Pleasure is the fundamental calorie that you want to be consuming as much of as possible."

By pleasure, he is talking about the warm, hopefully fabulous feeling of having good sex, which is one of several basic ingredients of any healthy and long-lasting relationship. Hookup sex can be a mixed bag. A study led by Paula England, PhD, a professor of sociology at New York University, called "Accounting for Women's Orgasm and Sexual Enjoyment in College Hookups and Relationships," found that 11 percent of women had an orgasm during first-encounter casual sex. That number grew to 16 percent for second or third hookups with the same partner.

When I arrived at *Cosmo*—already married with two kids—I was amazed at how many women depended on the magazine for basic sex ed. The number one question we got from readers, other than how to negotiate a pay raise, was, "How do I have an orgasm?" In the US, that lesson is not on any curriculum for sex education in either middle or high schools. It's why we made it an entire episode of *The Bold Type*—the show inspired by my experiences at *Cosmo*—on Freeform, where a writer is commissioned to craft a piece on orgasms only to fess up she hasn't ever experienced one. "The root of the problem is that no one teaches girls about pleasure in the context of sex education," says Marina Khidekel, founder of the newsletter *Undrrated* and now senior deputy editor of *Women's Health*, who edited *Cosmo*'s love and sex pages brilliantly for four years. "It's really about mechanics. This is what goes where. You learn about sperm and egg and condoms. You learn, 'Don't get an STD.' But then you are left on your own to figure out everything else." So we dedicated a lot of pages to just that: how to masturbate, fantasize, orgasm, and more. And we did it for the benefit of both sexes. "Whenever there was a story about blow jobs," Khidekel says, "there was another story about oral sex for women as well."

The goal was to get women to understand their bodies, what turns them on, and what brings them pleasure. Yet the statistics, as well as too many readers' experiences, suggest that sex is still often disappointing or alienating. There is a huge disconnect, and one that clear communication can bridge. "All of us want to be a really great lover, but we don't have the language that allows people to be vulnerable and awkward and unsure of themselves," Levkoff says. We strip naked, sometimes with total strangers, but don't tell them what we are hoping for in bed. Of course it is awkward! And it can be humiliating and frightening as well as ecstatic and awesome. The only way to solve it is to talk with the other person, which is more awkward still when you don't know them. "Instead of dealing with these uncomfortable spaces, we tend to rush to the end," says Levkoff. From the women's angle, that often ends with a faked orgasm—fast-forwarding to close a disappointing experience. But what is the point of pretending to have fun? No one wins at that game. "There is something really powerful about owning that awkward space, and saying, 'Look, I'm a little bit uncomfortable saying this, but I like when you do this . . .'" Levkoff says. "If you don't say what turns you on, then you don't really have the opportunity to get what you want in the end."

We also have to do away with what Khidekel calls the Disneyfication of the perfect partner: someone who will intuitively know what turns you on. "This feeling that when the right guy comes along, he will just know how to please me," she explains. "To leave that in their partner's hands—and not even know how to explore their own bodies and pleasure first—in this day and age, where women are so empowered in so many other areas of their life, shocked me."

Before you begin instructing your partner what feels good, you have to know it yourself.

Men have their first orgasms between the ages of fourteen and sixteen. Women run the gamut. "Some women may have had their first accidental orgasm at that age by rubbing up against a friend," Kerner says. "But most say, 'Oh, I didn't really have my first orgasm until I was twenty, twenty-five, or even later in life.'"

Figuring out what turns you on and gets you off is basic nutrition for anyone. So instead of getting out your journal, run a bath or climb in bed or lie down on the couch and experiment with getting to know yourself. While Levkoff is a fan of vibrators and other sex toys, she thinks it is important to start with your own hand. "Ask yourself, 'Are you comfortable enough with your own body to touch yourself?'" she says. "If not, then there are some bigger self-esteem body issues going on."

Exploring your own body is the only way to figure out what brings you pleasure. And that is the key to climaxing. "Being able to experience pleasure is very important to being able to experience desire," Kerner says. "Which is key to having meaningful sexual experiences that result hopefully in gratification and orgasm and leave you feeling connected and loved and tended to by your partner and also leave you wanting to have more sex."

This is the exercise portion of your love diet, akin to doing squats, stomach crunches, and plank poses (but way more fun). Add to this workout "psychogenic arousal," which Kerner defines as the ability to "fantasize and conjure up arousing, erotic scenarios, being open to erotic stimuli—which includes erotic literature or films." Kerner and Levkoff are both fans of what they call "ethical porn" (see Rule #10). "Fantasy raises your levels of arousal, which can be louder than the anxiety that many women feel around sex," Kerner explains. "So for a woman who is worried 'How does my body look?' or 'Are my roommates going to walk through the door?' being able to fantasize is going to be very helpful to generate the arousal and lower the volume on that noise."

Knowing how to prime your own pump is the pathway to pleasure. "Unlike male desire, pleasure really precedes desire for women," Kerner adds. "Whereas a guy can have a sexual thought, see or remember something sexy, and it can trigger the arousal platform. That's called spontaneous desire." Knowing that helps control the outcome as well because it is much easier for a guy to get off than a woman. This is not about how to please your guy in bed; it's about how to make the most of the fact that he is more likely to get aroused more quickly than you. On that note, we don't give men enough credit; most do want to please you in bed. But how will they know what that entails if you don't tell them? "They have no interest in someone faking it," Levkoff says. "They want to learn, and have as much trouble speaking up as we have trouble asking."

If you are too embarrassed to say, "I would really love it if you did this," then just show them. "Move his hand, or your body," Levkoff says. The way he responds is a good litmus test for a quality partner. If someone doesn't care, then they're not worth having in the first place. Why waste your time?

And so, before committing to have sex with anyone, ask yourself a few questions:

Why are you sleeping with this particular person?

Do you just want to have sex?

Are you sleeping with him because you want him to like you and you think the act of sex will achieve that goal?

Do you think it will be a good experience?

Once you identify your expectations, I suggest another speed bump (and potential buzzkill) in the form of a few more questions:

What is the potential upside?

The downside?

Is the equation worth it?

If this person has passed these first hurdles—and you may already be undressed and making out—this is where you can look for important cues:

Does he make you feel comfortable and safe?

Is he sensitive to the boundaries of consent?

Can you talk about sexual protection?

PS: Protected sex is a prerequisite to experiencing pleasure. As Kerner says, "Very few women who have unprotected sex feel pleasure because they have some level of anxiety about the sex that they're having."

Once you feel safe, then you can take it further. Tell him what feels good or what you like. And do you care about making him feel good, too?

"I would look for the ability to be able to tell a partner, 'Ooh, that hurts a little bit' or 'a little lighter' or 'a little softer please' and feel that there was a feedback loop happening," Kerner says. If it is not, and you are growing concerned, your body will ultimately switch from a pleasure-seeking mode to one that is on alert. "The body is always assessing the environment for feelings of safety and comfort," according to Garcia. "If you feel safe or you're having fun or you don't have fear, you go down a different biological pathway. If you're in a frozen state versus a mobilized, fun, safe-feeling state, you have a different biological response."

The goal is always for sex to be glorious fun, but given the stats of foreplay morphing into sexual assault, you need to think ahead to how to extricate yourself if things get uncomfortable.

As Steve Kardian suggested in Rule #5, it's okay to arm yourself with excuses. Pretend that you suddenly feel sick to your stomach (few people relish being vomited on) or that you have to go to the bathroom. Don't hesitate to say you have an STD. Especially if you never want to see him again. Say your period just started. Say whatever you need to say to get out of the situation safely.

Unfortunately, we know that it may not be enough to stop certain men. I wish there were a surefire way to spot those who mean to harm you. Or the ones who won't respect you. Think of the world of pain we could all avoid if such a test existed. There are signs (and we detail them in Rule #11), but there is no better place than in bed to know if a partner is worth keeping. Does he truly care about you? Can you trust him? Only then can he or she bring you pleasure.

TO DO

"It's so hard for women to know their sexual selves," says Wednesday Martin, PhD, the author of a forthcoming book on female infidelity called *Untrue: Why Nearly Everything We Believe About Women, Lust, and Infidelity Is Wrong and How the New Science Can Set Us Free.* "There's so much white noise and clutter and interference. Empower yourself."

In short, figure out what turns you on. Use your hand, a vibrator, the dryer on mega spin, whatever works. Once you know what gets you going, then practice having that conversation with yourself—so you can be articulate with any potential partner. While we are at it, let's also dismantle the myth that women are not as sexual as men. A 2014 study published in *Current Sexual Health Reports* found that women crave sex as much as men do. Martin describes women as "omnivorous" when it comes to sex. "Women crave variety and novelty," Martin says. "Our sexuality is the place where we have most been conditioned to believe that we have to be a certain way, and it is the place where we most need to assert our autonomy."

Knowing that you are a truly sexual being then exploring what that means specifically for you is the first step toward finding true pleasure. Then being able to share that self-knowledge with a partner means he or she can be an integral part of the experience. That is what we are going for here!

CASE STUDIES

Sara*, 21, on dating mostly men but learning about true pleasure from a woman.

Sara identifies as queer. "I date mostly men but have had girlfriends, too," the college senior says. "I like the openness of that label." She's also the daughter of four mothers—her biological mother was in a committed relationship with a woman when Sara was born. They split two years after her brother was born and have since each found new partners, so she has two mothers and two stepmothers as well. "Growing up in such a gay-friendly family allowed me to accept sexual fluidity as a norm," she explains. "Although it took me until college to really delve into my queerness, I was lucky to never feel confined by heteronormative pressures."

She grew up in Brooklyn, where being the black daughter of four mothers did not stand out quite as much as it did when she started at a small liberal arts college in upstate New York. At the time, Sara was still involved with her high school boyfriend. "I was considered off-limits freshman year," she says. "It was a long-distance relationship, so I would not hook up with guys, but I did wind up making out with my best friend." She says it happened consistently, and alcohol was always involved, but it never went beyond kissing. Her boyfriend knew, she adds, but did not care "because she was a girl."

Still, Sara felt a certain amount of curiosity watching all her friends use Tinder to hook up. So when she and her boyfriend split, she downloaded the app to see what she had been missing. "When I got matches, it was a confidence boost," she says. "But then I did not feel safe meeting these people, or I'd do some social media research and they did not look anything like their Tinder photo." She did hook up with one guy, whom she really liked. She was about to text him the next day to say so, until her friend advised against it. "I learned that there are rules to navigating hookup culture," she says. "If you text your feelings too early, then you are seen as being too clingy. My friend said it is better to distance yourself, which is hard for me. I'm a straightforward, touchy-feely person. That was frustrating."

In the end, she decided she had not missed out on anything. "The whole process felt dehumanizing to me," she says. Then she saw a video online that claimed black women have the lowest success rate "digi" dating. She was not surprised: "There's a lot of fetishizing going on, with Asian women, too." But in the end, she was simply turned off by how transactional online dating seemed. "It really feels like people are just interested in sex," she says.

Sara wanted the sex to be good if she was going to have it. "The sex with my boyfriend was great, but he had to learn my body," she says. "That takes time." The few hookups she had were more like "a lot of fumbling around," she adds. Oral sex with random guys, she maintains, is the worst. "Why do they need to stick their penis in my mouth?" she asks. "Especially if they won't go down on me?"

Then she started hanging out with a female friend who had also been in a long-term relationship with a guy. "I thought we were just becoming good friends—and then she kissed me one night," Sara recalls. "I was like, 'Hmm. What is this?'" What started slowly turned into a full-blown romance. "The relationship started with our really getting to know each other before we even started kissing," she says. "The opposite of hookup culture!"

The other thrill of this relationship was that she learned about what gave her pleasure. "I believe that girls should orgasm every time they have sex. When I had a boyfriend, sex was over when he was finished." Sara's orgasms were not the priority. But then she started sleeping with other women—and now makes them one. "Every single time I would get with a girl, I would orgasm," she says.

As a result, she has figured out what makes sex great for her. "It boils down to intimacy and trust," she says. "So I feel safe with you, and you care about my needs." To achieve that, Sara adds, she needs what she calls "the two *C*'s": communication and consistency. "I want to be able to say, I'm just as horny as you are!" she adds. "But let's talk about it so we are on the same page."

Leah, 24, on why she wrote her college thesis on how much hookup culture sucks.

Leah was tired of reading all the hype around hookup culture. "A lot of people who write about it haven't experienced it," she explains. "I couldn't find one piece where someone was being honest about what was going on."

Leah herself felt awful after consistent "pseudo relationships" at Middlebury College, where she'd hook up with the same guy for weeks, but he'd refuse to admit feelings, commit to something more, or show affection in public. However, it was one experience, with a guy who cut her off to explore other options after they had been hooking up for months, that made her think critically about it. "Months later, we wound up on a summer trip together and became really good friends," she says. After that trip, he asked her to be in a relationship with him, which utterly surprised her. She wondered what had changed and asked him. His response? He told her that he didn't think of her as a "human being" when they were hooking up.

That admission rattled her and inspired her to focus her senior thesis on hookup culture because she couldn't believe that people would actively participate in an experience where "people were not treating one another like human beings."

She was not alone: Of the seventy-five men and women she interviewed, she found only a few students who genuinely wanted a noncommittal hookup and nothing more.

Leah finished her thesis in 2015 and wrote a hugely popular essay for *Quartz* in 2016 based on her findings, called "A Lot of Women Don't Enjoy Hookup Culture—So Why Do We Force Ourselves to Participate?"

Since then, Leah has thought a lot about the pressure to join in on something that makes so many women, as well as men, feel bad. "There's less cultural pressure to be in a monogamous relationship like there was twenty to thirty years ago," she says. "The postcollege narrative is to be independent and feminist, which is inconsistent with 'I want to find a stable, committed partner who is also a man.'" This lays the groundwork for a culture that publicizes sex with no strings attached. The problem, she adds, is that "especially in college, few people are thinking critically about whether or not it works for her. It comes down to this consciousness."

Leah is still astonished by the number of messages she gets from women and men who have read her *Quartz* essay and related deeply to it. "Many say, 'I wish I could be as honest with myself as you are in this piece. I wish I could admit that I'm not having orgasms or admit that I just want somebody to love me,'" she says. "These are such basic things."

"Emotional attachment makes for better sex, so even if all you're in it for is sex, then at least have good sex."

So is being honest and vulnerable—two key ingredients to any serious relationship, yet almost impossible to be when hooking up. "Pretending that you're not going to have feelings about the person you are having sex with is just not realistic," she adds. "Emotional attachment makes for better sex, so even if all you're in it for is sex, then at least have good sex."

Now twenty-four and a journalist, Leah has been in a committed relationship since her senior year with a guy who was at first apprehensive about what that meant, exactly. "He struggled with 'If I commit to you, then what if I want to hook up with other people?'" Leah says. "Other avoidance versions I heard were 'If I commit to this person, what happens when we go to different places for internships? Or graduate and move to different cities?'" Leah had an answer for her now boyfriend—and for everyone else who worries about these things: Who cares? Live and love in the moment. "I told him that I had feelings for him and did not have time to wait for him to figure out how he felt about me," she says.

Then she laid it out for him. "I said, 'Best-case scenario, we stay in a relationship and we're really happy,'" she says. "'Worst-case scenario, we break up and maybe we dislike each other and that's okay, too.'" Her point being that "there are plenty of people who dislike their exes—but you learn something from the experience, so just take the chance."

Rule #10

Porn is like chewing gum— all artificial flavor.

After ten minutes of watching porn, I want to go home and screw.
After twenty minutes, I never want to screw again as long as I live.

—ERICA JONG

One evening in LA last year, over a couple of drinks, a male friend, a successful Hollywood producer who dates a lot of women, told me that he never ejaculates inside them, only on their faces. One FWB he sleeps with regularly dislikes this so much, he added, that when he's near climaxing, she fishes under her pillow for an airline eye mask, which she quickly slips on so she doesn't have to watch his finale. When I asked him why he persisted when it's so clear she doesn't like it, he claimed he was terrified to get her or any woman pregnant. He denied it, but I'm assuming he's living out his porn fantasy because it's so prevalent; men are always cumming on women's faces, often into their eyes. Which seems as pretty clear a metaphor as you can find.

At *Cosmo* the face issue was a question we got frequently: "My boyfriend wants to cum on my face. How can I tell him I don't like this without hurting his feelings or making him think I'm old-fashioned?" Props once more to the fearless advertising legend Cindy Gallop, who turned her intense dislike of the facial money shot into a TED Talk insisting, "Don't do it! We don't like it!"

Porn sex is not real sex.

"Porn is to sex what McDonald's is to food," says Gail Dines, a professor of sociology and women's studies at Wheelock College in Boston and author of *Pornland: How Porn Has Hijacked Our Sexuality*. "Stripped down to its most base elements: empty calories, high fat. You take a really healthy, fun, creative desire to eat or to have sex and in its place you give it a robotized, industrial product that basically is going to harm you."

To get even more granular, porn is to sex what Chicken McNuggets are to an organic free-range roasted chicken.

I will let the academics argue about whether porn leads to more or less sexual violence. My biggest problem with it is that so much of it is overwhelmingly degrading and humiliating to women. It's women being slapped about, flipped over like fish, and rammed in every orifice with huge hairless penises by men shouting "Whore!" or "Slut!" and then, the biggest lie of all, the women pretending they can't get enough.

Porn is to sex what Chicken McNuggets are to an organic free-range roasted chicken.

One of the most depressing things that I heard persistently from young women while I was at both *Marie Claire* and *Cosmo* was their resignation that men now expected them to behave and look like porn stars. Pubic topiary is the most obvious example, with total bikini waxes now the norm and the lasering of all pubic hair increasingly common. We had readers who told us that they had sex with young men who, having had their default sex education on Pornhub, assumed actual pubic hair on women was abnormal or even worse, freakish.

It's depressing to have one's body image determined by the clichés of the porn industry. But my real objection is that it limits people's expectations and imaginations about sex.

Back in 2014 Tracy Clark-Flory wrote a terrific piece for *Cosmo*, one that touched a nerve—no pun intended—and generated a ton of reaction, titled "How I Stopped Acting Like a Porn Star and Had the Best Sex of My Life." Like many millennials, porn was Clark-Flory's primary source of sex education and that of her partners, too. And so throughout her twenties, she pretended that it turned her on to have partners declaim about what they were going to do to her with their "big monster cocks."

After a particularly vigorous romp fit for a YouPorn pop-up ad, "One partner told me, 'I feel like I just discovered *Playboy* for the first time!' But *I* never, ever orgasmed," she confessed. "The only pleasure I got from these sexcapades was the satisfaction of feeling desired." Adding insult to injury, her constant depilating led to razor burn so painful it hurt even to wear the lightest of panties.

The big reveal for her was that "sex did not get sexy until I stopped acting like I was Jenna Jameson . . . and got real about what actually turned me on." That happened thanks to a new boyfriend who made her feel sufficiently confident to experiment with what excited her. As she explained, "I wanted something more from sex, beyond just making the guy feel good."

When I was growing up, porn was available in magazines like *Playboy* and *Penthouse*, the depiction of sex positively prim compared to what is available today online. VCRs next brought pornographic tapes into the privacy of people's homes—and in the eighties and nineties, porn videos were rented four times more than regular Hollywood films. And then came the internet. "It made porn affordable, accessible, and anonymous," says Dines. "The three *A*'s that drive demand. Porn sites now get more visits per month than Amazon, Netflix, and Twitter combined."

Pornhub, one of the most popular porn-sharing sites, got 2.6 million visitors per *hour* in 2016. In Pornhub's exhaustively thorough global analytics, referred to as the Kinsey Report of our time in a 2017 *New York* magazine cover story, the US was hands down the largest consumer of porn, with Iceland a distant second. Users are 75 percent male in the US, 25 percent female. The age data starts at eighteen and stretches well past sixty-five (where "hot granny" and "hairy pussies" are popular searches).

That they don't include any data on the under-eighteen group is problematic, as many studies have found that curious young boys coming into their own sexual awakening type "sex" or "porn" into a Google search and can quickly get barraged by violent images. Two clicks away and up pops "gagonmycockslut" and "roughassblaster." There's a porn genre for every niche. As I write this, Pokémon porn is trending.

And yet, as Erica Jong, creator of the zipless fuck concept in her groundbreaking novel *Fear of Flying*, points out, porn can provide a shortcut to getting turned on. Forty million people in the US alone regularly visit porn sites. Dines explains: "The job of porn is to get a guy hard and get him off in seven to eight minutes. So yes, it works in that specific way. But if we want a society where you connect sex to love and intimacy, then it absolutely doesn't work."

Good sex is so much more complicated than the language

of porn. There's no longing or buildup in the vast majority of porn being produced today—it is all immediate gratification, which is not how real relationships work. As to the effects of porn on relationships, a study published in the *Journal of Social and Clinical Psychology* in 2012 found that porn consumption lowered commitment in both men and women and had a stronger impact on men. Another study found that porn-free relationships are stronger and have a lower rate of infidelity.

The behavioral psychologist Mary Aiken calls online porn the "largest unregulated social experiment of all time," specifically because it is so widely accessible to children and teenagers. Gail Dines argues that exposure to porn is actually altering the "sexual template" for men who grow up on it. "Men have literally lost the capacity for intimacy and love or even know what connection is, by virtue of being brought up and socialized through porn," she says. And that, of course, has a direct impact on healthy relationships as these young boys become men whose default understanding of sexual relationships is through the lens of porn.

"Sexual trauma has become the new normal," says Dines, who lectures on porn's impact on intimacy and love on college campuses nationwide and globally. Both men and women pack her audiences, and she sees how porn impacts both terribly. "These guys do not want to be using porn this way," she says. "They feel ashamed. And the last thing you want around sexuality is shame—it's a killer for healthy sex."

If I were to swap a food pyramid for a love pyramid, respect would be your fruit and vegetables.

Healthy sex is, again, a key ingredient for any love relationship, and like everything else in this diet, it circles back
to mutual respect. If I were to swap a food pyramid for a love
pyramid, respect would be your fruit and vegetables. Those are
the essential nutrients without which it's hard to stay strong and
build a viable immune system. So if porn is something you and
your partner both like and you can use it in a way that helps you
get satisfaction and aids your sexual communication, then all
power to you. But using it as a guide for how to find pleasure or
intimacy is like eating at Taco Bell 24/7 and assuming this is as
good as it gets.

Porn could not be further from the real work and intimacy
of a relationship because you are a spectator, a voyeur, not a participant, watching other people's scripted, artificial, and often
violent sex lives. The only way you are going to truly figure out
what excites you is by first asking yourself, then experimenting,
and then communicating with your partner.

For those who like being voyeurs, Cindy Gallop started her
website MakeLoveNotPorn to get people talking about sex. "The
problem is not porn," she insists. "The problem is that we don't
talk about sex in the real world." On her site, you can choose
from a curated selection of "real-world sex" films made by regular
people (and a few porn "stars" who wanted to have real-life sex
on film). They have rules—films must be consensual, contextu-
alized, and porn-cliché-free. They are also open to every kind of
sexual identity and experience as long as it is respectful. "Our
tagline on MakeLoveNotPorn is 'Pro-sex. Pro-porn. Pro-knowing
the difference,'" she says. "And our mission is to help make it easier for the world to talk about sex openly and honestly—both in
the public domain as well as in intimate relationships."

I don't watch Gallop's MakeLoveNotPorn videos, but I do

love her argument that good sex is borne out of good communication. "I readily ask people, 'What are your sexual values?'" says Gallop. "And no one can ever answer me because we're not taught to think that way."

Leaving the issues with the porn industry and the way it treats its female stars aside—for those interested I recommend the documentary *Hot Girls Wanted*, but be warned—it's depressing. Gallop likes pointing out that there is porn out there that is not violent or degrading to women. "I get enormously frustrated when people use the word *porn* like it's one big homogenized mess," she says. "That's like using the word *literature* and saying it's all the same thing. It's as rich and infinitely varied as for the genres, subgenres, types, divisions, etc. of literature." Pornhub's number-one searched term in 2016 was "lesbian porn," followed by "stepmom" and "MILF." Other terms trending at the time of publication include "hijab porn," "squirt," and "gangbang."

I appreciate Gallop's pro-sex mission. "Since we don't talk about sex, it's an area of rampant insecurity for every single one of us," Gallop says. "We all get very vulnerable when we get naked. Sexual ego is very fragile. People therefore find it bizarrely difficult to talk about sex with the people they're actually having sex with, while they're actually having it." This circles back to knowing what brings you pleasure first and then being able to communicate that to a partner whom you trust.

"Sexual ego is very fragile. People therefore find it bizarrely difficult to talk about sex with the people they're actually having sex with."

"Many of us, if we're fortunate, are born into families and environments where our parents bring us up to have good manners, a work ethic, a sense of responsibility, accountability," Gallop says. "Nobody ever brings us up to behave well in bed. But they should, because empathy, generosity, sensitivity, kindness, and honesty are as important there as they are in every other area of our lives."

TO DO

I had never even considered the term *sexual values* before Gallop raised it, but it's a great idea to figure out what yours are.

Write them down, as a continuation of what sexual pleasure means for you, from Rule #9, as the two are interconnected. Your values extend to what you want from a partner as, in my opinion, the best sex makes sure both parties are having fun.

For that to happen, you need to home in on what that means for you:

Are you someone who prefers to make love in a bed beneath the sheets?

Or are you an exhibitionist, and having covert sex in public places turns you on?

Do you like to experiment?

Do you wish you could experiment?

Do you like men/women? Both?

What do you need to feel safe?

What do you need to feel alive?

Think of the best sex you have ever—or never—had. Write down that scenario.

Think of the worst sex you have ever had—and write that down, too.

Now you know what to seek out and what to avoid.

You may never need to think about these scenarios again. But they are useful to reflect on now in the event you come across someone who either makes you feel uncomfortable in the moment, asking for something you don't want to do, or someone who takes you beyond your wildest dreams and to the moon and back.

And last but not least, there is no need to act like a porn star, unless of course you want to.

CASE STUDIES

Grace*, 21, on realizing "being like a porn star in bed" was not going to get her the boyfriend she wanted.

Grace learned everything she knows about sex from watching porn. "I first watched it when I was a teenager," the college senior says. "The first guy I had sex with texted me afterward and said, 'You have sex like a porn star!' I took it as a compliment, like I was doing it really well."

This was precisely why she watched porn. "I based everything I did with guys on what I saw in porn," she says. And while the different guys she hooked up with seemed to love her moves, she quickly realized it was not all that fun—for her. "The first time a guy came on my face was traumatic," she explains. "It was disgusting—I literally could not see. You never see that perspective in porn. They just cut to the man's face! Not to the woman, who has semen in her hair and eyes." There was also the time when a guy asked if he could "slap her" as they were having sex. "That was directly from porn," she says. As were all the times when blow jobs made her retch, literally. "The sound of gagging turns guys on," she says. "Also, from porn."

Grace did not enjoy any of it but thought that's what it meant to be "good at sex," and now she wonders if that's why guys always wanted to hook up with her but never date her. "I'm the token single girl among my friends," she says. Grace started hooking up in high school and has continued in college. While the sex can be "great," she allows, it's the aftermath that she finds so depressing—as well as confusing. She spent the first semester of her sophomore year hooking up with a guy from her friend group. "Whenever I saw him out and about, he wouldn't even say hi to me. But then would text me at 1:00 a.m. every Friday and Saturday." This move is, she says, "ridiculously classic." That relationship finally faded, and she wound up hooking up with someone off-campus, which was the best hookup of her life. "I didn't have to stress about seeing him in the dining hall the next day," she says. "The waking up afterward, thinking, 'That was great.' But then you freak out because you wonder, 'Is he ever going to speak to me again?'"

The only worse feeling is when there was too much alcohol in the mix. "People get sloppy and you don't know the person so that means less communication," she explains. "I've definitely been in situations where I wanted to use a condom and we were both drunk and I did not communicate my wants clearly enough." Those feelings of worry and remorse, Grace adds, are far worse than "Will he talk to me afterward?" "Women are the ones who deal with the ramifications of unsafe sex," she points out. "Not the guys."

Her porn habits have shifted over the years. "I still watch it, but not the gagging or violent kind," she says. "I'm more interested in the women. I have a set type I find attractive. But I only watch it alone—never with a partner."

This, she admits, can also be depressing. "I've always wanted a serious relationship," she says. "I'd rather be with someone than watch." She has certainly tried to find that, and spent a semester hooking up with someone she felt very serious about. "When it was time to make the move to a real relationship, he was not into it. I was devastated." Even worse, the final breakup was through a text.

That same pattern set in again her junior year. But this time, she was going to brunch and making dinners with the guy she was hooking up with, which she considered dating. After several months, she told him that she wanted to take their relationship seriously. "He said that he only wanted to keep hooking up with me and to fuck as many girls as he could before he graduated," she says.

Grace ended that relationship and took a break from sex and dating altogether. "I realized that I was always leaving situations feeling disappointed and broken down," she says. "Until I meet someone who treats me well, I don't want to fill the void with men who don't."

The break has allowed her to reflect on her own role in these relationships—as well as the impact of porn on her love life. As a black woman who has had white men say, "You're pretty for a black girl," or "You're the first black girl I have ever been with," she became interested in how interracial porn shifts a viewer's image of race and sex, and she is writing her college thesis on it. "Looking back I cringe at the things I'd do to get boys to like me," she confesses. "I was making an active choice to let men do degrading things, whether cum on my face or ignore me, because I wanted someone in that moment. But in the end I realize that the only way anyone is going to respect me is if I respect myself first."

Jess, 28, on replacing shame with forgiveness in order to save her marriage.

Jess was nineteen when she met Vaughn, just six months after she had given birth to her first child. "He lived across the street and I helped him and his brother push his car out of the snow," she recalls. They discovered that they went to the same Church of Jesus Christ of Latter-day Saints temple, and Vaughn invited Jess back to his house one afternoon with friends after church. "Everyone left and we found ourselves talking until 6:00 a.m. the next morning," she says. "We didn't kiss or anything—but we never ran out of things to talk about."

Jess even told him that night that she had just placed her

daughter up for adoption. "He didn't even bat an eye," she says. "He just said, 'That's cool.'"

That was a first for Jess, who had had several one-night stands with random guys in the months following her baby's birth. "That wasn't like me," she says. "I was acting out in my grief." Most men were not so accepting. "When they learned I had a child, they either treated me like I was really easy to get into bed, or like I was damaged goods," she says. Vaughn was different. "He's very grounded," she says. "I just felt this sense of peace and calmness around him."

A week after they met, she invited him to her daughter's adoption finalization. "I was standing outside the LDS temple sobbing, and he just held me," she says. "He was twenty-two at the time and just knew that I needed someone to hold me." That was the moment when Jess knew she was going to marry Vaughn.

But then she borrowed his computer, and his search history popped up. "That's how I learned that he watched a ton of online porn. I was overcome with all these awful feelings—betrayal and disgust. But mostly I felt like there must be something wrong with me if he needs to look for that elsewhere." She confronted him, and he promised he would stop. Jess was not all that experienced, either, so she just trusted him when he said he would stop watching it. But it took a toll on her self-esteem. "I was suddenly self-conscious in bed and out," she admits. "I felt I couldn't compete."

Vaughn assured her that was not the case and that he no longer looked at porn. She believed him, and little by little let her guard back down. Ten months after they met, Vaughn proposed and Jess said yes.

But then right before their wedding, she borrowed his computer again and it was a horrible déjà vu. He hadn't stopped. The couple sought counseling from their church, which is where Jess learned porn addiction is rampant in the LDS community because "there's so little talk about sex," she says. "That's often how boys get their sex education, which is totally twisted." They were referred to an outside counselor and learned that in order to address Vaughn's porn problem, they first had to learn how to communicate with each other. "He grew up in a family of yellers," she says. "And my family sweeps it under the rug."

That was a long road. They did get married, and she discovered porn was still a problem right after their first daughter was born. "That was such a painful time," she says. "I was confused why these fictional gross caricatures of humans were so much better than me. But I was also completely distraught over my post-pregnancy body." She became repulsed by the thought of having sex with him—and seriously considered divorce.

After several nights of talking, during which Jess saw and remembered the man she first fell in love with, she thought about how he'd held her as she sobbed outside the adoption finalization, and she thought, "Am I really willing to give up that easily?"

Vaughn had a similar epiphany. "He realized that his family was way more important than cheap thrills," Jess says. And she realized she had to be more open and honest about sex in their relationship. "A huge reason why porn addiction is so rampant is because there's so much shame attached to sex," Jess realized. "I was like, 'You know what? I need to show him there isn't shame in having issues, there's just shame in not finding a solution for them.'"

Jess decided to help him instead of leaving him. "We agreed to nightly check-ins, where we talked about any urges he had that day," she says. "We also put filters on our computers and phones." The two worked on the plan together as well as on their own to spend more time focusing on their own love life. "I realized that even though I had two kids, I was still so naive about sex. Our sex has just gotten better with time because we talk to each other! I can say, 'I like it when you do this!' Or, 'Do you like this? Let's try this.'"

As much as porn almost ruined their marriage, it became the catalyst for them to create true communication around sex.

Rule #11

Stick to natural sugars. Substitutes are bad for your health.

A healthy relationship gives us energy and nourishment. But when it goes badly, it can be devastating. This rule is short but important as it addresses one of the most difficult issues in dating. One of the reasons I wanted to write this book was to address the following statistics:

One in three women in the US has been physically abused by a domestic partner according to the National Coalition Against Domestic Violence.

One in five women has been raped. And almost half the time by a man she knows.

Love can be dangerous. And while this rule is shorter than most, it is in many ways the most important as I realized that there is no modern and concise primer on how to find a healthy relationship or how to avoid a dangerous or toxic one.

Similarly, there are no relationship development courses in the school curriculum. We spend more time researching what type of car we should buy than we do on the people we get naked and share bodily fluids with.

So here are some pointers to bear in mind before you embark on a new relationship, because abusers and sexual predators don't wear T-shirts that say, "Avoid me at all costs!" On the contrary, they are often handsome and charismatic and sweep you off your feet.

In his book *Food Rules*, Michael Pollan recommended reading the labels and making sure all the ingredients were not only natural but easy to pronounce.

When it comes to relationships, there is one "label" all women should do their best to avoid: narcissistic personality disorder. It is difficult to detect because it masquerades as a delicious sweetener, but it comes with a bitter aftertaste.

"The most common psychological characteristic of all abusers is narcissism," says David Adams, an expert on domestic violence and the cofounder of Emerge, the oldest abuser intervention program in the US. Narcissistic personality disorder is defined in the *DSM-5* (the *Diagnostic and Statistical Manual of Mental Disorders*) as displaying "a pervasive pattern of grandiosity, need for admiration, and lack of empathy." That can manifest specifically in the following ways: "a grandiose sense of self-importance," "preoccupation with fantasies of unlimited success, power, brilliance or ideals of love," and "requires excessive admiration."

There are more signs, such as the individual or person "has a sense of entitlement" and "is interpersonally exploitative." But perhaps the most chilling is this sign: "lacks empathy: is unwilling to recognize or identify with the feelings and needs of others."

Empathy, the ability to put yourself in someone else's shoes, is a key ingredient in any relationship.

Interestingly, while women can be abusive, too, narcissistic personality disorder affects men more greatly than women. The term *Prince Charming* has often been used to describe abusers because at first they present as terribly romantic and superattentive, and likely lavish you with compliments. To anyone who has not been in a relationship or had sex in months, or who is vulnerable and just getting over a breakup, this can feel too good to be true. And it is with abusers, Adams warns. "Elaborate, overly romantic gestures on the first date could be a warning sign," he says. "He brings you flowers, or insists on paying for the meal." Adams calls this "entrapment style" and recalls a victim of an abusive relationship he interviewed for a court hearing. In describing the relationship to Adams, she said that on the first date she told the man who became her boyfriend that her phone was broken. On their second date, he presented her with a brand-new phone. The two became a couple—and she became one of the one in four women who experience abusive relationships.

It is easy to fall for what seems generous at first. How kind! How thoughtful! But it's a clear sign of too much too soon. So is oversharing intimate information early on in the relationship. Look for boundaries and respectful talk about ex-girlfriends. "Does he rag on his ex?" Adams asks. "Or say things like, 'I can tell that you understand me more than she ever did.'"

Any talk of "soul mates" on a first or second date also makes Adams cringe. "It is a false intimacy. Some women respond to this idea that he understands me or he gets me—but that doesn't happen in one or two dates. It takes time."

Not sharing information or outright lying are also red flags, Adams says. He said he owned a business when he did not. Or he said he had a house, but it was actually his mother's house. These are two examples Adams uses as "predictive of lying and manipulating, which is abusive behavior."

And then there is misrepresentation. "For instance, you discover that something he has said is not true, like he is more involved with his ex than he says he is," Adams says. "So then the question is, How does he deal with that when you confront him?" Does he flip out and ask, "How dare you question me?" Or does his reasoning make sense to you?

You have to trust your gut here (more on this in Rule #12). This is not an easy one to decipher. But if you have a sense he is lying or hiding something from you, do not brush it off. This brings to mind the term *gaslighting*, which Adams has written about as a form of "psychological manipulation, also known as 'crazy-making' or 'mind-messing,' whereby the victims are made to doubt their perceptions of reality, and ultimately, their sanity." That feeling—that you are the crazy one—is another earlier indicator of an unhealthy relationship.

A more concrete warning sign is insistence on early sex. "First date sex is incredibly common among abusers," Adams says. What is particularly tricky about this red flag is that many women have sex on the first date with men who wind up being wonderful partners. But if he happens to be a narcissist with a predilection for abusive behavior, then it is serious. "For abusers, sex signifies ownership," Adams says.

This chapter may be freaking you out. Take a deep breath. If you see any indication that a man you are interested in or seeing has abusive tendencies, Adams suggests testing your theory by setting a boundary and seeing how he reacts. For instance, he wants to see you on a night when you have other plans. You have a work event. Or dinner plans with friends. "If he makes an issue of it, that's a good test," he says. "He may even say, 'I feel like I care about you more than you care about me,' or make you feel like you have to make a choice."

Establishing boundaries early on is a good test of any relationship. Is this person accepting of the other demands in your life? If not, you should reconsider. If you are not sure, then start paying close attention to your interactions and don't brush anything under the rug.

Track your dating data. Some clues: If you feel in any way that there is a disproportionate balance of power early on in the relationship, think it through carefully. Maybe it's something you can live with or maybe you sense it will recalibrate in your favor over time. For example, he may be older and further along in his career, but you'll catch up. If you're in doubt, here are some questions to ask yourself:

How does this person make you feel?

Did he swoop in and sweep you off your feet?

Did the swoopy bit last as long as you had hoped?

Did he say he loved you within the first week?

Are you worried about how he will react if you say you can't do something with him?

Does the expression "walking on eggshells" now apply?

Do you feel responsible for his feelings?

A yes to that last question is the biggest warning sign to extricate yourself. "It will only get harder," Adams says. "The longer you are in the relationship, from his perspective, the more you are his."

If you have doubts, don't silence them. This is really important. Despite the success of Bumble, women still expect men to set the pace. Don't be passive. Be active and always listen to your doubts. Is he getting too close too quickly? Is he suffocating you, competing with your friends or parents for your time?

Try not to let the relief of meeting someone half-decent who does not seem to have baggage or a crazy Facebook-stalking ex or kids cloud your judgment. The pyramid gets smaller as we get older because people pair off. So who is left? Why are they left?

Does he talk endlessly about himself?

Does he bad-mouth his ex?

Does he keep asking how much he means to you and then setting up tests where you have to reinforce that he comes first?

Does he criticize your friends? Does he compete for your attention around them?

Does he get jealous of any other people in your life?

Yes to any of the above is an indicator of a narcissist. Dr. Jean Twenge's fascinating research specializing in millennials focuses on the uptick of narcissistic traits in a generation she says grew up being told they were special. They turned up for games and got a trophy for participation; their grades were inflated; their parents applauded every cough, spit-up, and burp. Her book *Generation Me* argues that social media exacerbates all that early parental indulgence. "All the recent changes in our culture, specifically social media and its impact, encourage a cultural individualism," Twenge explains. "So there is more focus on the self and less on social roles and the collective in general." This can be good—more positive self-views, she says, and "a trend toward more equality around race and gender." But it can also lead to greater levels of narcissism. "It's where the self is all-important and other people are only useful for what they can do for you."

Asked how narcissism affects relationships, Twenge agrees it's a train wreck. "Narcissists are very charming," she says. "But then you learn that you are in love with a person who does not actually care about you. That is a very bad formula." She adds that the cultural messaging that demands you have to love yourself before you can love someone else is a disaster. "Not really!" Twenge says. "People who really love themselves are called narcissists. They make horrible relationship partners."

And it goes both ways.

While men are more likely to be narcissists, plenty of women fall into that category as well. "Our culture's focus on individualism says you don't need anyone else to make you happy—you can make yourself happy," Twenge says. "You don't need relationships and if you think you do then you are a weak person. That is one of the messages that is coming across in our culture today. And besides, there are all these choices out there so there is no need to choose one now and settle down. Where that hits a wall is, it's a natural human tendency to need relationships."

You are reading this book because you want a relationship. And the point is to find one that will fill you up with happiness and sustain you in the long run. One that takes work—they all do—but is rewarding because you get what you give.

CASE STUDIES

Leanne*, 26, on when "too good to be true" was in fact that.

After Leanne graduated from university, she quickly worked her way up to becoming the manager of a high-end restaurant in Boston. She was making great money by her twenty-third birthday, loved her job, and had a terrific group of friends, but found meeting potential boyfriends particularly tricky. "I get off work at 2:00 a.m.," she says. "Not the best time to go meet someone for the first time."

For this reason, she relied on internet dating because she could look for possible partners on her own clock, but she met a bunch of duds interested only in sex, and one psycho who, after she rebuffed him online, started sending her rape fantasy notes. She blocked him, took a break, and then went back to online dating with more caution the next time around. That's when she met Sam.

The two met for coffee, and at first, he seemed almost too good to be true. "Handsome, charming, and so smart," she recalls. "The only odd thing was that he did not work." She agreed to a second date, and a third, and learned that he was independently wealthy. He did not need to work, which meant he could flex with her schedule, an added plus. "I introduced him to my family a few weeks into our relationship," she recalls. "My dad was like, 'I've never seen you so happy!'"

But then she introduced him to an old boyfriend, who was working with Leanne on a business idea. Following the meeting, Sam flipped out. "He became irate and jealous," she says.

"He even suggested buying my ex out of the partnership, which was so bananas that I brushed it off."

Suddenly, Sam started showing up unannounced at her apartment, or at work, which Leanne found off-putting. He said he wanted to surprise her, but she felt as if he was checking on her. Still, she wanted the relationship to work, so she batted those thoughts away and focused on those early impressions.

And then, a few months later, the relationship finally began to unravel. "We were grocery shopping when he got a call from his grandmother," she recalls. When he got off the phone, she could tell that something was terribly wrong—but instead of telling Leanne what had just happened, he stormed off.

A few weeks later, and following more erratic behavior, Sam finally admitted that he was bipolar. His grandmother had learned that he had stopped seeing his doctor and taking his medication right after he started dating Leanne. "I was devastated he had not told me this earlier," she says. "And yet, I was in love with him by then—and he was sick. I wanted to try to work it out."

It got worse and finally culminated in a manic meltdown three months later. "A friend's visit set him off into such a rage, he started smashing furniture in my apartment and then punched a brick wall and shattered his hand in the process," she recalls. "I finally got him out of my apartment, and as I was bawling in my apartment, I was like, 'How did I let this happen?'"

Leanne ended the relationship quickly—and safely. Still, it rattled her. "I was lucky I got out early, before anyone got really hurt," she realizes. After a six-month hiatus, she went back online, this time with a new perspective: "I am taking it way more slowly. And heeding any unsettling signs."

Mary*, 38, on wanting to get it right the second time around.

Mary met Sebastian at a party in Manhattan. He was an artist from Chile, with thick black curls, deep brown eyes, and a strong accent that she found irresistible. On an early date, he brought her to the top of the Empire State Building. On their way back to his SoHo loft that same night, he literally swept her off her feet—picking her up and twirling her around like a scene from a movie. "No one had ever treated me that way," she says. "I felt like a princess."

But then he took her to a party the following week to meet some of his friends. One was a rowdy Irishman who grabbed her hand and dragged her onto the dance floor. Mary was having a blast, dancing to the B-52s, when she noticed Sebastian sitting in the corner of the room, glaring at her. "I went over to see if he was okay," she says. "And he literally turned his head and refused to talk to me."

They left the party, and the silent treatment continued. Finally, after she begged him to tell her what she had done to upset him, he shouted, "How dare you dance with my friend!" She was stunned. "I'd never been with a jealous person before," she explains. "It startled me." She spent the next few days apologizing—even though she knew she had done nothing wrong. "I just wanted it to go back to the way it was in the beginning," she says.

Finally, he returned to his romantic self, though Mary made a point of never dancing with, or even being overly friendly to, another man. She even found herself lying about whom she went to lunch with—giving the name of a female colleague instead of a male. She once panicked when she wound up on a work trip with a man. "He was gay, but I worried Sebastian would freak out. So I lied and said I was traveling on my own."

This was all within the first few months of their relationship. When it was just the two of them, everything was fine, but never as fun as it had been in those early days. When he proposed to her out of the blue at a restaurant one evening, she told him she needed to think about it. "I knew by the look on his face that was not the right answer," she says.

By then, she was practically living with him, and the silent treatment that began that night lasted two entire days. "At one point, I was literally weeping, begging forgiveness," she recalls. He finally relented and again things went back to normal. A month later, when he proposed again, she said yes.

Days before the wedding, Mary's mother took her aside and asked, "Are you sure you are doing the right thing?" By then, Sebastian's family had flown in from Chile for the wedding, the dress had been altered, and the flowers ordered. Even though she knew, in the pit of her stomach, that this would not be an easy relationship, she told her mother she loved him. "I will never forget the look in her eyes," Mary says. "It was a combination of pain and defeat." Mary's father was emotionally abusive, and that marriage had ended in heartbreak after her mother discovered he had been having an affair. But Mary did not want to think of patterns, or daddy issues; she was getting married.

She does think about them now, ten years later and in the wake of her divorce. Sebastian's jealousy grew and morphed into other controlling behavior during their marriage. His anger was explosive and terrifying. And while he never hit her, he'd say, "You're so lucky I don't hit you."

Ironically, Sebastian left *her* after seven years of marriage. "It was over children," she says. "I wanted them, he did not." And like everything else in their marriage, she thought she could work it out. "When he said it was over, I was stunned," she says. "And then totally relieved." She was also thirty-six and wanted to be a mother. After a year hiatus, she started online dating, but first she made a list of what she wanted in a partner. "Kindness was at the top of my list," she says. "And someone who either had kids already or wanted to be a parent with me."

She went on at least a dozen awkward, dull, or plain awful dates. "I made the mistake of telling one guy I liked that I wanted children on the first date," she says. "He literally asked for the check." So she decided not to bring kids up until the fifth date at the earliest and started to despair that she may not ever make it that far with anyone. "There was one guy who was dressed head to toe in black spandex," she remembers with a laugh. "And only ate orange food."

She'd almost given up when she was set up on a blind date with Jim. Their first dinner lasted four hours and ended with him walking her home and asking if he could see her again. On the second date, at a Patti Smith concert, he admitted sheepishly that he was "earnest." "I told him it was my favorite quality in a person," she says. And then on the third date, she asked why he and his ex broke up. He explained that she did not want children, and he did. "At that point, I said, 'I was going to wait until the fifth date to tell you I want kids, too,'" she says. "And I added, 'PS, I hope we make it to the fifth date!'"

They not only made it, but he proposed three months later. By then she was certain that he was the kindest man she had ever met. They married the next year and wound up becoming parents through adoption. "During one of our last fights, I remember Sebastian saying that he couldn't believe that he had wasted seven years of his life with me," she says. "I was thirty-six at the time and knew my chances of getting pregnant were getting slimmer with age." Meanwhile, after two heartbreaking miscarriages and a subsequent diagnosis of age-related infertility, Jim told her that he was open to adoption. "He said, 'I don't need to have a biological child—I just want to be a parent with you,'" she recalls. "It was the most earnest—and romantic—thing he has ever said."

PART FOUR

LOVE RULES:

THE SLOW DIET

FOR THE LONG HAUL.

Rule #12

Trust your gut and protect yourself with probiotics.

It sounds obvious, but you should date people who make you feel good about yourself—not anxious, or as if you need to improve on a few areas of your character. Yes, it really is that straightforward.

Trust your gut:

Does he listen to you? Respect your feelings? Make you feel beautiful? Or does he comment on your weight in a negative way? Interrupt you in the middle of a sentence? Make you feel that if you just try harder, you could do better? Per my previous rule, he doesn't have to be a total narcissist; he can still be a jerk. Is he homophobic? Or racist? Either would knock him off my list. Do these things matter to you? Finding the right match means fitting this puzzle piece of a person into your larger life.

Are you a devout Catholic and he is an avowed atheist? Can you and he have a civil, engaged conversation about your divergent beliefs?

All these details are important, as too often women ignore the off-color remarks or the off-putting tics because he is good looking. Charming. Funny. Or great in bed. He can be all those things—and still not be the right match for you.

This is where you have to really home in on what makes

you not only happy but also feel as the French say *bien dans sa peau*—good in your skin. This is a person you want to share your life with, and clear, open communication is key.

So if he makes a joke that makes you queasy, do you feel comfortable pointing it out and explaining why it bothered you? Are you worried about his reaction?

As with everything else in this diet, I am asking you to be conscious and proactive.

If you have met someone who is a potential long-termer, notice all the ways he or she makes you feel—when it is great or when it is odd. Write them all down on your pro-and-con list. If "I don't like his haircut" or "He doesn't have a coffee maker" are the cons, you can work with it. If they are "He makes fun of my religion" or "He gets irate when he is drunk, and that is often," you might reconsider.

Recording potential problems will help you see the pattern clearly. (This is a good time to check the intolerance list you made in Rule #3.)

If you find yourself once again attracted to the unavailable ones or the superhandsome but not terribly smart ones or the too-cool ones or the mama's boys . . . pause.

And before you go any further, think of what makes a good partner.

The one ingredient that you must consider is kindness, as it is the basic foundation of any relationship. And, thrillingly, it is also a love probiotic that you can produce yourself.

Study after study reveals that kindness is key to a healthy and happy life. One study published in *Psychological Science*, titled "How Positive Emotions Build Physical Health," found that "positive emotions, positive social connections and physical health influence one another in a self-sustaining upward-spiral dynamic." Another study, out of the University of California, Berkeley, found that witnessing acts of kindness produces oxytocin, occasionally referred to as the "love hormone," which aids in lowering blood pressure and improving our overall heart health. Emory University also did a study on the effects of being kind, finding that when you perform an act of kindness, the same areas light up in your brain as if you were receiving kindness or experiencing pleasure. The phenomenon is known as the "helper's high."

This is the most fundamental and straightforward rule in this diet: Seek out kind partners, and be kind yourself.

So he likes to paint his face and shout from the stands during football games. Does he listen to you? Treat you well? Open doors for elderly women? Stop to pet dogs in the park?

This is the most fundamental and straightforward rule in this diet: Seek out kind partners, and be kind yourself.

Nice is good for the long haul, especially in this modern age, where women's working lives matter as much as men's. We used to take the focus on men's prospects for granted—they were the breadwinners. Their careers mattered most. That is no longer the case, so it is time to flip the script, or at least split it down the middle. This is a dialogue, not a monologue. So ask yourself, "Is this someone I can have a conversation with about both my career ambitions and my parenthood ambitions?" The assumption used to be that even if you were a career girl, you would take time off from your job to have the children, that your job would be secondary to his. That is shifting. Women are not giving up their jobs the way they used to.

How does that impact the conversation you have and your relationship?

So look at the checklist of what has not worked in the past, and make a new one for what might in the future. And make sure kindness tops it. Fill the rest of it out based on emotional well-being—forget the superficial stuff—and even then, remember to be flexible. It is like going to the farmers' market and discovering that they don't have mozzarella but they do have feta. You can adapt. You can find flexibility and a willingness to try something new but related to what you want and need. It may wind up being even more delicious—as long as that key ingredient, kindness, is involved.

And yet, our culture has us hungering for the low-fat blue cheese–flavored dressing that does not need refrigeration because it's so full of chemicals. The cool guys who play hard to get. The players who flirt with you on Friday and then ignore you on Saturday.

This is one of the reasons Whitney Wolfe started Bumble. "We're trained to be attracted to meanness," Wolfe points out. "Watch any Disney TV show—the little boy is mean to the little girl because he has a crazy crush on her. But then he goes home and writes love letters at night." Or the cliché fairy tale of a young woman falling in love with the beast because only her love can transform him into the great prince. Wolfe wants to rewrite that story. "We've all heard, 'Oh, he's too nice,' as if it is a bad thing," Wolfe says. "Why is that part of our culture? You'd only be so lucky to end up with someone who's too nice. That would be a beautiful life."

She's right. Nice and kind do not have to equate to boring.

Nice is not white bread, milquetoast. It is curious and interested. It also isn't threatened by a woman's ability or intelligence. Women work as much as men; we need more help raising kids as well as running the household. And that is where kindness becomes really important—as well as an acceptance that your career matters as much as his does. If you are a career woman, then your partner must have an ability to see you as a full person and not merely an appendage to his life or someone who aids him in his journey. He must see you as a copilot.

As important as it is to ask yourself what you want in a partner, it is equally important to ask what you think he or she wants. Someone who is committed to a clean fifty-fifty balance? Or someone who wants to put their career on hold to watch the kids while you work? Someone who insists you are the one to leave your career, and they are the sole breadwinner? And if that changed, over the course of the relationship, are they flexible enough to change as you each develop? Are you?

What are the things you need for the long run? What about your partner? What is it in you that they have spotted, however subconsciously, that they feel they need in a partner?

Pay attention. Be honest. Is this a relationship between equals? Look for any huge disparities. Call them out before you commit.

And it goes without saying, the ability to have a conversation is another good indicator.

Whatever relationship you are in, there has to be a mutual respect. This means you must feel that you are able to voice your opinions, your concerns, your points of view freely. And you must be able to listen to his. The minute you stop caring what each other thinks, you have to move on.

Some relationships last fifty years or more—others don't.

But the length does not matter as much as the quality of the relationship while you are in it.

CASE STUDIES

Beatrice*, 36, on knowing whom she wanted to marry at 23.

Beatrice says that she figured out what she wanted in a relationship when she was ten years old. "My parents divorced when I was one and remarried other people when I was four," she explains. "I was the flower girl in both of their weddings." Both sets of parents divorced again when she was heading into ninth grade. By then, she had two younger half sisters and a half brother, and saw firsthand "what unstable relationships can do not only to the people who are in them, but to their children." Meanwhile, she grew up in the film business in Los Angeles, where both of her parents lead seemingly "fabulous" lives. "My mother is an actress, my father a producer, and my stepfather a writer and director," she explains. "There were a ton of distractions and a multitude of reasons to get divorced."

Beatrice says her mom and dad were great parents, but not the best relationship role models. "I wanted the opposite of that," she explains. After dating only four people in "her entire life," she met Tim in college. Postgraduation, Tim moved to San Francisco, and Beatrice to New York, and the two subsequently split for six months. During that period, Beatrice tried dating and learned both that it was "really hard" and that she was "not very good at it." Meanwhile, Tim had begun sending her long, heartfelt letters campaigning to get back together. "I realized then, this man loved me and had expressed every desire to become the person that I felt I needed in a partner," she recalls. "Sexy people come in and out of your life, but I was focused on sustainability." She was twenty-three when they got back together and twenty-seven when they got married.

By then, she had begun a career as a movie producer, and although counterintuitive, she decided to start her family, which was incredibly challenging. "We were married for three years when I gave birth to my first child," she recalls. "So I was producing a film and breastfeeding." Her daughter was three months old when the lead actress dropped out, and since all the funding was based on her participation, Beatrice had to work around the clock to keep the film from falling apart. "I was coming home at midnight, and my breast milk dried up from the stress," she says. "It was insanity." It also put a strain on her marriage. "Tim was like, 'We have a child now and you are acting like this movie is more important than us!'" Beatrice says. "I did not want to hear that at the time. By the way, it's a terrible movie and no one's ever seen it." She took his point to heart.

They now have three children, and Beatrice has gone on to produce award-winning movies. Keeping her family a priority has been a good ballast against getting swept up in the adrenaline rush of the movie business. It has also made her fall more deeply in love with her husband over the years. "I actually think that Tim is a better person than I am," she says. "Not to demean myself, but he is kind. He wakes up every Sunday and takes our kids to church and lets me stay home. He's funny, supersmart, and has an ability to see ten steps ahead, which is not something I have." What she brings to the relationship is a gung ho attitude, enthusiasm, and a deep belief in him, and them as a couple. "He recently started his own private equity firm, and I said, 'Do it! And if it doesn't work, we'll sell our house and move into a rental,'" she says. "Failure doesn't scare me so much. My attitude is, 'We can make it work.'"

As both are full-time working parents, their life is truly equitable, another reason why Beatrice thinks their relationship works so well. This approach was first tested when they had children. "We called it Evie Stevie. If I woke up with Rosie early one morning, then he would do it the next. Or if I had to deal with her in the middle of the night for two hours, then he'd take her to the park the next day. It was a conscious effort to even the playing field again."

It worked. In the end, Beatrice believes, their relationship is strong because of that attention to equity. "We're able to say what is on our minds," she says. "And work it out." Looking back over their nine-year marriage, she admits now that her choice was not "purely romantic or heart-driven. I acted with my head and then my heart followed. In a funny way, I'm more in love with Tim today than I was when I was nineteen."

Cristina, on testing her professional matchmaking skills in making her own match.

Cristina founded the Los Angeles–based Matchmakers in the City with her sister Alessandra in 2012 as an antidote to online dating. "Busy professionals work hard to succeed in every area of their lives, but most have put dating on the back burner," says Cristina, who prefers not to reveal her age. "They know the value of their time and want to avoid wasting it on those with the wrong intentions. Or they don't want to risk seeing their interns on the same dating app." Committed to helping people find an "old-fashioned love story," the sisters have grown their matchmaking business to reach beyond Los Angeles, to New York City, San Francisco, and Washington, DC, as well. So she knew she was in good hands when Alessandra introduced her to Dan. "She screened him for me," Cristina says.

Cristina also had her own rigorous vetting process. "I wanted someone who shared my faith," she says. "That was important."

Alessandra knew Dan from the confirmation class they both taught at their church, and she thought he was a good match for Cristina. She introduced the two at a friend's birthday party. The next time they saw each other, Dan asked Cristina to dinner. She accepted, and that's where she laid out for him her approach to dating. "I told him that I wanted to build a friendship first," she says. She also wanted him to pursue her, and made that clear. "I said, 'I won't reach out to you, but if you want to call me, you can,'" she recalls. "And then I added that if he wanted to see me, that would be great, but the next date had to be with a group."

She realizes that it was a lot for him to consider, yet she believes that women need to set more clear parameters and boundaries when it comes to relationships. "There are no barriers to dating anymore," she says. "It used to be your dad at the door. Now, women are in college, in high-power jobs, and living on their own. It's wonderful, but that becomes problematic for dating since you have to erect your own boundaries, otherwise you can be taken advantage of very easily."

Cristina coaches the women she works with to create

boundaries first simply by making a list of internal qualities they're looking for in a partner. She did the same for herself. "Reliability was number one. A personal relationship with God was number two. And while I believe you have to be attracted to your partner, it has nothing to do with the way he looks. I wanted a good man," she says. Another requirement was an ability to dance. "The second time that we saw each other, we were dancing with friends. He's six five and actually a good dancer, but did not care about looking cool. He was having fun."

After a few more outings, Cristina realized that she was being too rigid about "group dates only." "My sister and I say that every relationship has a diamond and a setting," she says. "Someone who sparkles, and the other who embraces that." Cristina is the more outgoing of the two, whereas Dan was someone who "did not need the limelight." That made getting to know him in group settings difficult. For their next date, she suggested that they go out to dinner, and then they alternated between one-on-one dates and gatherings with friends until three months later, when they officially became girlfriend and boyfriend.

Cristina says she knew from the minute she met him that he might be her future husband, but she took a year to test her theory. "I tell our bachelorettes that it is important to see this person in all four seasons," she adds. She told Dan the same thing.

Still, she was surprised when he did propose, a year after they first met. "I was on my way to the same Adoration service that I had gone to earlier on the night that we met, and invited him to join me," she says. "It was a last-minute thing, so when he asked me to marry him, there, at the service, I was caught off guard."

Even though it was a surprise, she had already rehearsed her answer: yes.

Rule #13

Set your own "best before" date.

By now you know your weak spots and cravings. Whom you fall for, what types you should avoid, as well as whom and what to strive for.

Armed with that knowledge, now is the time to construct a timeline. How old are you? And what do you want? By what age? Whether you want to find a boyfriend or girlfriend before your next birthday, meet the person whom you will marry before you are thirty and have your first kid by thirty-one, or find a new partner in your fifties or sixties to spend the rest of your life with, make a map of where you are right now and where you ultimately want to be. You can, and will, make detours—we are all human after all—but it helps to identify where you are starting from and ultimately where you are planning to get to.

Especially since the old adage "You spend your twenties trying not to get pregnant, and your thirties trying *to* get pregnant" remains the truth (and oh, the irony!) for so many women as peak fertility is around age twenty-six, which feels quite early for anyone just getting going in the workforce.

The good news is that there are now many ways to become a mother, and it's worth thinking about what you are willing to consider ahead of time. I know many women in their midthirties hoping they will find a partner to have a kid with—and as many women in their forties and fifties who feel cheated out of motherhood because they assumed they had more fertile years than they did. Several interviewed for this book say that remaining childless was one of the hardest things they have had to face. Others said that they are so relieved that we live in a time where they can admit they don't want children—and find partners who don't want them, either.

When Cameron Diaz announced that she was "never drawn to being a mother" for a story in *Esquire*, she sparked a cascade of stories from women of all ages who were relieved to have a successful movie star say out loud what they were feeling. Writer Sezín Koehler compiled for *HuffPost* a list of eight reasons she decided not to have children that included sleep, money, and interest. The thirty-five-year-old added, "Between my creative work, a job I love, and a husband I adore who agrees with everything in this article, I am happy, healthy, and the most fulfilled I've been in my life." She also said, "I don't need to push a real child out of my vagina to be a woman. And I don't need a child in order to be happy."

Polly Vernon, an editor at the *Guardian*, wrote a great piece on being child-free in which she confided that she knew she did not want kids from the age of seven. "By my midteens," she wrote, "I could quote statistics on the damage that kids would wreak on my career trajectory, finances, social life, and body. Although to be honest, none of that was, or is, as big a factor in my decision to remain childless as my instinctive feeling that I just didn't want children." In the end, she finds the cultural obsession with having kids "boring." She, like Koehler, found a partner who agrees with her. "[He] has never questioned my enduring childlessness," Vernon adds, "and . . . I love [him] all the more as a consequence." All this is true for her—and I am glad we are at a moment where women can admit these things so freely.

So what is true for you? Know this yourself first and then know that this conversation must be had early on in any prospective relationship. Not on the first date, of course. But certainly as you get serious. This is not something you want to raise for the first time on your thirty-sixth birthday, having silently nursed the hope for six years that you're on the same page only to discover that you have different ideas where kids are concerned. I know too many women who assumed they could nudge a partner in the direction they wanted without ever having asked directly and were then devastated to learn that that person was either not ready or not interested in having children. Or even worse, totally interested in having kids, just not with them.

Tanya Selvaratnam's book *The Big Lie: Motherhood, Feminism, and the Reality of the Biological Clock* should be required reading for women who know they want a family. "The big lie is that we can do our things on our own timetable," Selvaratnam explains. "We are taught and conditioned to think proactively about our work futures, but not about our personal family future." An actor and artist, she speaks and writes from personal experience. After marrying at thirty-six, she had three miscarriages before she turned thirty-eight. "The experience of having that first miscarriage felt so privately painful," Selvaratnam says. "I didn't realize that they are so incredibly common. My partner at the time did not find out that his own mother had had a miscarriage until I had one!" She started in vitro fertilization (IVF) only to discover during the process that she had gastrointestinal cancer. Her infertility may have saved her life—but also strained her marriage, a well-documented phenomenon of the impact of infertility among married couples.

Now forty-two, divorced, and still without a child, she published her book to rave reviews and lots of attention, but it did not sell as many copies as she had hoped. Granted, infertility and illness are perhaps not the most obvious ingredients for a bestseller, though that didn't stop *When Breath Becomes Air* or *Being Mortal* from taking off. But young women can find it especially hard to confront their desire to have children for fear of not seeming serious about work or of being accused of harboring retro Ozzie-and-Harriet fantasies.

Sylvia Ann Hewlett also learned this the hard way in 2002 when she published *Creating a Life*, which panicked women by pointing out a stat hiding in plain sight: 40 percent of women earning over $50,000 were childless at the age of forty-five. One of the first books to talk honestly about the incongruity between women's fertility and women's ambitions, it sold very few copies in comparison to the amount of media attention it received, which suggested that women did not want to face the painful truth about fertility: that the older you are, the harder it is to get pregnant, no matter how successful you may be. Hewlett—who has five children, the youngest conceived when she was fifty after years of trying—was stunned by the response. "In my reporting, I stumbled on the fact that the more successful you were as a woman, the less likely you were to have a partner and children," she explains. "Whereas for men, it's the mirror image: the more successful you are, the more likely you are to have at least one if not more wives, and several children."

While it may not have become a bestseller, Hewlett's book did alert women to their fertility fallibility. It's now much more widely understood that fertility drops as women age. The American Congress of Obstetricians and Gynecologists states that a woman's fertility gradually begins to decrease at thirty-two and then more rapidly at thirty-seven, which coincides with the number of eggs a woman has throughout her lifetime: from 300,000 and 400,000 at our first menstrual cycle to somewhere between 39,000 and 52,000 by age thirty. By forty, that decreases to between 9,000 and 12,000, and not all are viable. And so women who put off childbearing in their more fertile twenties to focus on their careers need to know that it is harder to get pregnant in your less-fertile thirties. "I call it a creeping non-choice," says Hewlett, referring to the fact that 30 percent of women who are either on track in their careers or have a degree don't have children at age forty. "I don't like the word *childless* as it denotes a deficit, and not all women feel that way," she adds. "But I can say that career success leads to all kinds of great choices for men. For women, those choices are less generous."

The booming billion-dollar fertility industry has also enticed women to believe that we can still drink in the last-chance saloon by turning to tech to become mothers. IVF has become— for women who can afford it—a fallback. It has created the assumption that even if you started trying late and nothing's yet happened, you can always make it happen with IVF. Yes, you can try, but there are no guarantees. And just because it's starting to get lots of sexy press coverage, don't hold your breath for in vitro gametogenesis, the latest development in the infertility field, which creates eggs or sperm from "pluripotent" stem cells found in a person's skin, male or female. Yes, this super sci-fi, and ethically challenging, procedure has been successful in mice—but it has a very long road ahead before it's a solution for humans.

Just as age is a factor for getting pregnant without assistance, it is also the most important factor influencing the success of IVF, according to the Society for Assisted Reproductive

Technology. Just under 40 percent of IVF cycles result in babies for women ages thirty-two and younger. At age forty, that success rate is cut in half to less than 20 percent. I know one close friend in her late thirties who did five rounds of IVF—and spent upward of $250,000—and got nowhere. The toll—physical, mental, and financial—was excruciating.

And yet, the good news is that many children are now born this way. One of my oldest friends—having endured multiple miscarriages in her early forties with IVF—carried her first baby at fifty-one after the clinic where she had stored her remaining fertilized eggs wrote saying they would be destroyed within the year. With some distance on the grueling series of miscarriages, she felt she had nothing left to lose by having the last eggs implanted, and nine months later she gave birth to a gorgeous and healthy baby girl whom we still refer to as a miracle. But you should know that the science is still young, and though the clinic brochures are glossy, statistically the odds are against you. Only 40 percent of all IVF rounds in which the mother's biological egg is used will end in a live birth. So to be clear, 60 percent don't work.

The answer to the question, "Do you want kids?" will affect any relationship you have with a partner. . . . You must know where you stand on this issue—even if you are ambivalent—to avoid feeling sideswiped later.

While the emotional and physical toll is intense, the financial impact can be debilitating for many, too. Costs, which are not covered by most insurance companies, average $12,400 per woman, according to the American Society for Reproductive Medicine. One study found that 70 percent of all women who sought some form of assisted reproductive technology went into debt in the process.

So ask yourself, "What do I want?" The answer to the question, "Do you want kids?" will affect any relationship you have with a partner—he or she will want them, already have them, or have no interest in parenting whatsoever. You must know where you stand on this issue—even if you are ambivalent—to avoid feeling sideswiped later.

Dedicate space in your notebook to this subject, and start by answering the following:

Do you have kids?

If not, do you want them?

If so, how many? By what age?

What's your family fertility history?

And finally, how do you want to have them?

I ask this because there are so many ways to become a mother these days. And so Hewlett's description of the "creeping non-choice" of age-related infertility has been countered with a multitude of options. Adoption, for starters. Surrogacy. Sperm banks. Looking for someone who already has kids. Or how about finding a friend you want to have a kid with? This may sound radical, but it certainly takes the pressure off your love quest. "Fifty percent of people do not raise their children with the person they had the child with," says Esther Perel, who also suggests, "So why don't you pick a friend, with whom you can actually raise a child?"

In tech parlance, it's a disruptive idea, one that an Israeli organization called Alternative Parenting has pioneered by pairing single men and women, both gay and straight, who were ready to have a child but didn't have an intimate partner to do it with.

I love this proactive approach. The most joyful piece I ever commissioned when I was the editor of *Marie Claire* was a story called "I Had a Baby with My Gay Best Friends."

Kitty was thirty-three when her first marriage ended. By then, she was running a successful restaurant, had a wonderful group of friends, and felt ready to become a parent. Her next boyfriend, however, was not, and felt pressured by what to her felt like a ticktocking fertility time bomb. Fast-forward to thirty-eight, she implemented her own version of Alternative Parenting by asking her two gay best friends to be the fathers of her child. One provided the sperm; the other, his last name. Contracts were drawn up, and a turkey baster was used. Olive was born in 2009, and while their coparenting is not without its issues—the dads don't approve of refined sugar and a sometimes-exhausted Kitty has served Chef Boyardee for dinner—they all agree Olive is the best possible result. Kitty writes that the arrangement has also made dating easier: "Lots of women my age have children with exes they hate, so the fact that I love Olive's dads and they're involved in her life is refreshing to guys and takes some of the pressure off." She adds that working "Olive's dads" into first date conversations has led to a few hilarious double takes.

I love her problem-solving approach to this issue. I can't count the number of women who have said to me, "I just never imagined I would find myself in this position. I'm in my late thirties, I really want kids, and there is NO ONE around." Acknowledging there are alternate ways of becoming a mother can open up all sorts of possibilities.

Thirty years ago, most women were dependent on marrying someone who would support them financially. That is no longer necessarily the case, but no one will dispute that it's easier to have someone to help you care for a child.

Even if one had perfect control over it, figuring out when to have a family is one of the most complicated challenges for modern women now entering the workforce with such gusto. Work environments—especially big corporations, until recently mostly male conceived, dominated, and directed—have not built in enough supports around working motherhood. I would like to say "parenthood" here, as we crawl toward a day where the onus falls upon either parent, regardless of gender. But currently the dominant dialogue in the culture is fixated on women "juggling" or "balancing" or "having it all," which adds to this looming, dreadful sense that it is such a struggle to have a job and raise kids. Yes, that is true, but it is really difficult to have a job and be going through IVF, too. And it is also really hard to be happy at work if you left it too late to have kids and you really wanted them. We need better-paid family leave and family-flexible work policies. We also need more positive stories of women who are happy with their working-mom lives.

Elizabeth Gregory set out to do just that in her book *Ready: Why Women are Embracing the New Later Motherhood*, in which she interviewed more than one hundred women who had their first child, by birth or adoption, at or after age thirty-five. "It's not new for people to have children late in life," she says. "What is new is that they are having their first child later." She writes from experience, having had her first biological child at thirty-nine and having adopted her second child at forty-seven. She wrote her book as an antidote to all the scare-tactic stories urging women to have babies in their twenties. "It's always about how women should hurry up and have kids sooner than later or they will be sorry," she says. "They seem to sabotage women making advances."

Meanwhile, delayed parenting is women's response to the advances of birth control, as well as to the gains we have made in education and the workplace since that historic day in 1960 when the first contraceptive pill was approved by the Food and Drug Administration. Add improved public health to women's growing opportunities—which means more babies survive birth and people live much longer—and you see how women have responsibly adjusted to these changes. It's societal support that is still stuck in a patriarchal past. "Women today delay child-bearing until they can afford quality childcare," Gregory says. "There's nowhere to put children for the first five years, and then they get out of school at 2:30 p.m. and are off every summer. It's a social system set up to make women fail."

For this reason, Gregory warns against encouraging women to have children in their twenties unless they really do feel ready to become mothers. What Gregory also found through her research is that while older mothers may not have anticipated becoming mothers so late—and may have fewer children than they envisioned as a result—they do have more stable marriages, greater confidence, and financial security, which has a direct and positive effect on their children. She also shares the many ways in which women can become mothers later in life. "The oldest woman I spoke to was fifty-six when she started her family by adoption," Gregory says. "I also talked to people who had no problems, including one who had five kids between thirty-five and forty, and many women who went through unsuccessful fertility treatments before adopting."

I, too, waited until I felt ready to have kids, having assumed the best thing I could do was be financially stable and see enough of the world that I wouldn't resent being grounded with babies. I was thirty-six when our first son was born and thirty-nine when I squeaked in with number two. Given my odds, I still feel so grateful I managed to get pregnant with no fuss. But if I had my time again, I would do it earlier and have tried for three or even four. At thirty-six, I had no idea how tired I could be—I remember once being too tired to reach the bedroom and instead lying down on our hardwood floor and falling straight to sleep—or how much love I could feel, or how much worry I could take, but far more important, how much fun and how fascinating being a parent would be. That's because no one really tells you the good stuff anymore. Their father remarked one evening that having our boys was like going from living in black and white all his life to suddenly living in color. It was the most romantic thing he could have said.

Throughout the years, we have passed the baton of responsibility for them back and forth depending on who was traveling, working on deadline, or earning the most. I cannot tell you how many times I am asked by earnest young women, "How do you balance being a mother and holding a job?" I always say, "There is no balance. I don't even know what that would feel like. I don't mind extremes and there is only borderline chaos. You have to embrace the chaos."

So, if you want kids, then whatever age you are, own your fertility. Have a frank conversation with your gynecologist. We get our cars checked before long trips—top up the tires with air pressure, make sure the brake pads are not worn out, and change the oil. Treat your body the same way, because the road to becoming a parent—whatever route you take—can be winding and expensive. Be prepared.

IF YOU DON'T WANT CHILDREN

Know this is something to discuss with future partners as men and women have such strong feelings on the issue. Finding a partner who wants the same thing as you do when it comes to kids is key to having a healthy and fulfilling relationship. You don't want to be responsible for someone else's regret.

IF YOU ARE AMBIVALENT

Do your research and know there are lots of options down the road should you decide later in life you want to be a mother.

IF YOU WANT CHILDREN

Research your own fertility options. Talk to your mother and sisters and aunts. Ask about their ability to get pregnant, if any of them had a miscarriage. Research other issues surrounding fertility that may be hereditary.

Talk to your gynecologist about getting a fertility workup, including a basal antral follicle count, which tells you the number of eggs available for pregnancy, as well as a hysterosalpingogram, which is an X-ray of your uterus and fallopian tubes to determine their condition. "A woman in her thirties may have fewer eggs than another woman in her forties," Selvaratnam explains. "The experts themselves cannot explain how it works because every woman is different. Everyone has to be their own advocate and seek out the answers and information." On that note, for those women who think that they will get pregnant in the future because they have been pregnant in the past, know that your age is the number one indicator of your fertility prospects.

Consider egg freezing. Knowing that egg viability declines in your thirties, don't wait until you are thirty-eight. But if you do decide to do it in your twenties or even early to midthirties, take into account that it is expensive and can involve serious side effects, including hyperstimulation syndrome, which can cause rapid weight gain and painful bloating. Not terribly sexy if you are also dating. And then there is the psychological impact— what happens if that egg does not take when you do decide to fertilize? Or the ethical dilemma of what to do with those eggs if you wind up getting pregnant naturally?

The bottom line is that the longer you put off resolving the issue, the harder all options are: The older you get, the more difficult it is to conceive naturally, to conceive through IVF, to freeze your eggs, and even to adopt. Whenever I hear that someone is pregnant and older than forty-five, I am astonished at how often people marvel at the miracle of it. The truth is many women who get pregnant for the first time at the age of forty-five or above likely used egg donors. "People lie about it all the time," Selvaratnam says. "We have made it abnormal for women to acknowledge readily when they have not used their own bio-logical egg." This is one of the last taboos.

Frankly, I am sick of taboos around this topic. It is time to talk openly about all the possibilities around becoming a parent—or about knowing you don't want to be a mother— without any shame. "The best reaction I had to my book was when young women said they could finally have conversations with their partners about this," Selvaratnam adds. "People don't know how to talk about this stuff."

CASE STUDIES

Maria*, 49, on wanting two kids by the time she turned 35, not three at 33.

Maria was newly married and had just turned thirty when she and her husband, Stefano, decided they were ready to have a kid. "I wanted at least two by the time I was thirty-five," the now forty-eight-year-old marketing consultant says. "A boy and a girl, three years apart." Thus far, she had achieved all her other life goals—including a handsome Argentine husband and an MBA from Wharton. "My twenties were all about living abroad, working abroad, getting my MBA and my first real job," Maria says. "I met Stefano at twenty-one and knew he was my partner for life—but there was no talk about children until our early thirties."

She was so sure she'd get pregnant quickly that she called Lisa, her sister-in-law who was also just married, and the two made a pact to get pregnant together. "We wanted the cousins to be the same age," she says. Six months later, Lisa called to say that she was pregnant. "I felt a knot in my stomach," Maria says. "I was happy for her, but also slightly jealous as I was still not pregnant." Maria decided to "use her type A, MBA personality" to laser focus on that goal: She bought fertility books and a calendar, as well as a special thermometer that indicated the best time to have sex based on her body temperature. This put a strain on her marriage. "If Stefano could not have sex for whatever reason during my peak window, I'd get really upset," she recalls. And still, she did not get pregnant.

After a year, her ob-gyn started running tests and discovered that her uterus was webbed with adhesions, which blocked her fallopian tubes as well. "I had severe endometriosis," she says. "That explained why my periods were always so painful." She started hormone therapy to force menopause, which would help eradicate the adhesions that were causing her infertility. "I was thirty-one and having hot flashes—often in the middle of pitch meetings at work," she recalls. "And sex was really painful because I was so dry."

She was thankful that Stefano was both patient and supportive. "For the first time in my life, I felt like a failure," she says. "If he was anything but positive and loving throughout this period, I am not sure I could have done it." None of this was part of her fantasy of motherhood. The treatment worked—but she was still not getting pregnant. Stefano had a good job, which meant Maria could quit her job and add acupuncture, meditation, visualization, and even emotional free flow painting to her regimen. Still nothing. Lisa had already had her baby when Maria's ob-gyn finally referred her to a fertility clinic.

Any fantasy of an easy or natural pregnancy was replaced with blood tests and self-administered injections. "I remember getting a box of syringes, and my heart sinking," she recalls. "Stefano got really good at administering shots."

Then they got the first bill—$12,000. They had budgeted for this—but Maria got ovarian hyperstimulation syndrome. "Instead of producing twelve eggs, I produced thirty-three," she says. Her doctor warned that removing them—necessary for the

IVF—was perilous, as the cavities would fill with fluid and that could potentially lead to a blood clot. "I asked about the risk, and he said that a gynecologist's daughter had recently died that way," she recalls, but that the risk was probably less than being in a plane crash. She was given the choice: risk the removal or cancel the cycle. "I had been doing this for months, and it was incredibly expensive," she says. "But the question was, 'Is this worth dying for?'"

She and Stefano stayed up most of the night debating what to do and decided to have the eggs removed—as the chance of dying was so slim. They did not tell her parents. As warned, the cavities filled, and she left the hospital with a catheter emerging from her abdomen and attached to a bag clasped around her leg. By then, she was back at work at a boutique PR company in marketing, where she had to empty the bag throughout the day in the ladies' restroom and wear thick white tights to prevent blood clots. "No one at work knew what I was going through," she says. "It was incredibly stressful." Her doctor attempted to fertilize all thirty-three eggs that were retrieved, of which thirteen took. Of those, their doctor recommended they pick three to implant because he was worried, she says, "that they would not all make it."

Another late-night discussion: The couple decided to have three embryos implanted and freeze the remaining ten.

Finally pregnant, Maria was astonished to learn that all three "took." She was still lying on the exam table when her doctor suggested "reduction"—an issue neither she nor Stefano had ever heard of before then. "Carrying triplets to term was considered very high risk," she says. "The big decision was, 'Do we save the lives of two by killing one?'" Yet another sleepless night, and they both decided to keep all three.

And Maria, who at five feet eight weighed 120 pounds, went on a high-calorie diet and was put on bed rest when she hit 180 pounds. "My cervix could not hold the weight of three babies," she says. She had a cerclage—"they sewed my cervix where it was tearing," she explains—but that was only a temporary fix. The staples gave way when Maria was twenty-six weeks pregnant—she had an emergency C-section the first day the births were considered viable.

Each weighed less than two pounds and stayed in the neonatal intensive care unit, where machines literally kept them alive for the first few weeks. Maria spent the entire day, every day, at the hospital, and Stefano came immediately after work. "They looked like miniature monkeys wrapped in plastic tape and wires," she says. "Their arms and legs were the size of my fingers." They were finally released three months later to go home.

The months that followed were a blur—and then a letter arrived in the mail from the fertility clinic asking what Maria and Stefano wanted to do with the remaining frozen blastocysts. "We could donate them to science, or to an infertile couple—or discard them," she says. "Another quandary." They decided to donate them to science.

In the fall of 2017, Maria was asked to sit on a panel at the corporation where she worked as VP of marketing. "We started a Lean In chapter at work and the first panel was called 'Can We Have It All: Mothers Who Work,'" she recalls. When Maria announced that she had "triplets and a singleton" the audience of seventy-plus women gasped. Oscar*, her fourth son, was an accident. "I did not realize I was pregnant for three months," she says. Her triplets were five years old by then and learning to read in kindergarten. "I was still so traumatized by the first birth experience that I did not think I could handle another." Instead, Maria and Stefano got the experience of a scheduled cesarean with no complications. She agreed to sit on the panel because, "I never talk about how hard it was for me to become a mother and now I see what a disservice that is to all the young ambitious women, in their twenties, who likely want what I wanted: to be a mother who works!"

Maeve*, 54, on finally becoming a mother at 50.

Maeve met Sean on a blind date when she was thirty-eight. "He wasn't my type at all," she says. "I pictured myself with a metro-sexual architect or some lean college soccer player turned graphic designer." At the time, Maeve was working her dream job for a boutique photo representation agency with photographers in New York.

She had spent her early thirties living in San Francisco with someone who felt like her soul mate but was not the right fit. "He was a master craftsman who built gorgeous custom furniture and loved the art world and the outdoors as much as I do," she says. "But he couldn't seem to get his act together." She felt that she had to encourage him constantly and worried that being his constant cheerleader would both be exhausting and make her bitter. Maeve still loved him when she moved to New York for work, which eased the inevitable separation. By then, she was thirty-six.

She spent the next several years focused on her career and put her love life and motherhood on the back burner. One of six siblings, she knew she wanted to be a mom one day. But no one—not her own mother or her gynecologist—raised her age as an issue. Admittedly, she says, she was in denial about the realities of her biological clock.

And then she met Sean. "He was not my type at all," she says. "A big suburban New Jersey guy—with the biggest heart around." Sean pursued Maeve, and she realized that she did not need a partner who could go to galleries with her or wake up on a Sunday morning and say, "Let's go on a ten-mile hike." Sean was a "solid, great guy," she says, "the guy you want on the *Titanic*. Stable, loving, and always makes the right choice." She chose him for his honesty, his integrity, and his dependability. "I realized that all I needed was a good guy who loved me and who I loved. We could be very different people, but we shared common goals and values. And he was as enthusiastic and supportive about what made me happy as I was about what made him happy."

They dated for a few years before Maeve brought up children. "I wanted at least two kids," she says. She realized, at forty-one, that it was time sensitive and told Sean that she was not sure if she needed to be married, but she knew that she wanted to have a child with him.

At forty-two, she started trying to conceive. "Shockingly I got pregnant quickly. And Sean proposed." She was two months pregnant when she miscarried. "I was devastated, but still delusional," she says. "I thought since I did it once, I could do it again."

The two married in Ireland at a fly-fishing lodge when she was forty-five. "I was still fantasizing that I was superwoman and would get pregnant again," she adds. When she did not, they went down the path of fertility treatments, with several rounds of intrauterine insemination. When these did not work, the obvious next step was IVF. At that point, she says, "We had to look critically at the costs and the risks. It was clear that adoption was the best way forward."

By then, her sister brought home her first baby—Jessica, whom she had adopted in China as a single mother. "The second I met Jessica, my path was crystal clear," Maeve says. "Sean's sister was adopted domestically, so he was already there."

The two switched gears completely and decided to pursue an adoption in Ethiopia as well as in China since Maeve knew either could take a long time, and she wanted two children. They thought they would adopt first in Ethiopia because it on average took less time than China, but after three long years filling out paperwork and working toward that goal, their agency informed them they were pulling out of Ethiopia. "I have never cried so hard," Maeve recalls. "We were virtually at the finish line, and so it felt at that moment like a miscarriage at nine months." While they respected the reasons for the agency's decision, the news was devastating.

Thankfully, they were 90 percent done with the paperwork for the China adoption and were able to focus on that once the dust settled from the "Ethiopian bombshell," Maeve says. "We shifted gears, focused on China, and were swiftly matched with a child who was identified as ready for adoption."

Maeve and Sean flew to China to meet their son, Dylan, on

December 23, 2015. On December 24, it was official: Maeve was a mother. They flew back to New York on January 3, and six days later, Maeve turned fifty-one. "It truly was the best birthday ever," she says.

She was so thrilled that she started the paperwork on a second adoption in China. "I knew that due to my age, we wouldn't have the benefit of getting one child and then seeing how it was before deciding to have another one," she explains. Tim, Dylan's brother, joined the family right before Maeve turned fifty-four.

"Sean told me years after he met me that he knew we were going to wind up adopting," she says. "If he had told me then, I would have been shocked. Now I know, I picked the right partner."

Rachel, 44, on deciding not to wait around for someone to marry before becoming a mom.

Rachel had always assumed that her life would unfold "the way lives unfolded," she says. "I would find the person, have the wedding, bear the children." But the lawyer and entrepreneur who cofounded TheLi.st, an online network for successful women, and Change the Ratio, an earlier nonprofit, said that she was having such a great time in New York, focused on her career and living what she calls "the dream: an exciting, whirlwind, fun, wildly free, independent life," that she never thought about her age or its implications—until her midthirties, when people began to ask, "So don't you want to have kids?"

"I'd respond, 'Of course I want to have kids, why is that your default assumption?'" she recalls. "That was thirty-five, thirty-six, thirty-seven, and thirty-eight. By thirty-nine, I was starting to get a little bit tense." That's because she did not have a partner who made sense, and she felt that her fertility was "a ticking clock."

When she turned forty, she started dating an older man who had an almost grown child and did not want more. "I realized that dating him was tantamount to a decision I had to make," she recalls.

That was 2014, and one of the threads on TheLi.st was about egg freezing, which Rachel described as "incredibly robust and useful." Women were matter-of-factly sharing the particulars of their experience, including their treatment and doctors and how much it all cost. "I was on a business trip sitting on the edge of my hotel bed weeping silently as I read all these messages," she recalls. "I was totally confronted with the enormity of the conversation, which I had not begun with myself. If you are a woman in want of a reliable supply of sperm to even see if you can get pregnant, that actually takes some doing."

Rachel realized that getting pregnant in her early forties was an odds game. "Basically, I needed to have sex," she says. This was a challenge as she was, in her words, "very single." She had no plans to start sleeping around with men willy-nilly—and even less opportunity, considering she was about to spend a month putting on musicals at a children's summer camp. Not the greatest timing for a woman who wants to get pregnant! (Nor, as most parents would hope, for anyone going off to summer camp.) But it was just a month. She'd think about it when she got back.

Rachel went up to the camp, which she'd attended in her youth back in Canada. Everything was familiar, including another former counselor from her hometown of Toronto who had also returned to direct a play. He and Rachel hit it off, as two forty-somethings directing musicals at a children's summer camp might do. Fast-forward to the end of summer—the relationship didn't last, but it had left a souvenir: Rachel was pregnant.

Single and forty-one, the decision to have the baby was a no-brainer. According to Rachel, there was never any other option. "This kid was a miracle," she says. "I knew the challenges, I knew the odds. No one could plan for this. But I got lucky."

Two years later, Rachel is happily a single mom to a busy, happy toddler. Her daughter's father lives in Toronto but is involved in her life and visits frequently. They are one of many nontraditional families in their New York neighborhood. "I'm pretty normal," Rachel says. "When I was forty, I was afraid I would never get pregnant. I felt like an outlier. But I'm not. Not by a long shot."

Being a single mom has changed her life, Rachel says, and she has no regrets. "There's no way around it—parenthood is hard, and doing it alone is even harder," she says. "But I have an amazing support system, great friends, and family. And this is what I wanted."

Her advice to single women looking to have kids is not to wait. "Do as I say, not as I did," she jokes. "Think about what you want, and what you will want. I got lucky, but luck isn't a strategy. Being prepared is."

Rule #14

Look for relationship role models.

One evening my husband and I were out to dinner with a group of couples when one of them started arguing.

It looked as though it might escalate when the man suddenly said to the woman, "BIC," and they both calmed down.

"BIC?" my husband asked the man later. "What the hell did that mean?"

"Bollocking in car," the guy replied. "We can shout at each other in the car afterward, but we have a rule never to argue in public."

We borrowed that rule from them.

No one emerges well from arguing in front of others. Though we all know people who do it, it makes the people you're with feel unbearably awkward.

As a whole you don't read much about couples who get along well—and their quirky rules for surviving each other—for the simple reason that it doesn't sell. Conflict and drama sell. Which is why celebrity magazines love nothing better than to stalk celebrity marriages when they look as though they're in trouble. Certain weeklies have stayed afloat miserably trafficking in the pending collapse of Brangelina, in Ben Affleck's various relationships, and in Chris and Gwyneth's conscious uncoupling.

Similarly, novels don't focus on happy marriages because where's the tension? Master of them all, Leo Tolstoy pointed out: "All happy families resemble one another; each unhappy family is unhappy in its own way."

In other words, happy marriages are not sufficiently interesting to write about, whereas unhappy marriages make a good story. As an artist, he may be right, but it doesn't help us if we're looking for good examples to follow.

Yet in both our careers and our spiritual lives, we are specifically encouraged to seek out role models because we can learn so much by simply watching others.

A standard question at job interviews is "Who is your role model?" No one asks that on a date. Imagine if your date leaned forward and asked for your relationship role model. And yet, these mentors can be helpful. Just as financiers study their rivals' deals and lawyers pore over previous trials to learn argument and technique, studying the behavior of couples you admire is worthwhile. And relationship role models are especially useful if you didn't grow up with parents who liked each other or treated each other—or you—with respect.

Look for couples who support each other publicly, make each other laugh, celebrate each other's achievements, smile obligingly as their partner tells a story they have undoubtedly heard a dozen times before. Look at the couples you admire.

So find a couple of role models. In real life, not on TV, where the scriptwriter always gets the last word, though it's hard not to hope that Pam and Jim from *The Office* will make it to their golden wedding anniversary, joined by Tami and Eric from *Friday Night Lights*!

Look for couples who support each other publicly, make each other laugh, celebrate each other's achievements, smile obligingly as their partner tells a story they have undoubtedly heard a dozen times before. Look at the couples you admire.

Older couples who have weathered personal storms are often the fount of hugely helpful advice and have the distance from their earlier crises to tell you the truth about how they cleared hurdles. In many cases they may say it's a question of ignoring each other's weaknesses and focusing on their strengths. Of paying attention to what their partner likes to do and not judging them. Those who have clocked up the years usually want other people's relationships to work out for the long haul, too. They understand the benefits. There's such hard-won pride in a fiftieth wedding anniversary.

Playing tennis with a good player improves your game, and hanging out with functioning couples will give you a mirror to reflect on your own behavior as part of a couple.

ACTION PLAN

Get out your journal and make a list of the people whose relationships you admire.

The first question you should ask yourself is "Did I grow up with a good relationship role model?"

If your own parents are still together and are happy, then

you have a huge head start. Watch how they treat each other, how they look out for each other, how they have handled difficult situations over the years.

If there's no way you would hold your own parents up as role models—even if they are still together—what about the parents of your friends?

Think of those whose homes you enjoyed hanging out in as you grew up. What about relatives, an aunt or cousin? Or perhaps you have siblings whose relationships you admire. Ask them for their thoughts.

Throw hypothetical situations at them and ask for their solutions. Instead of going to the movies, have dinner one evening and ask them for their top ten rules for making love last. Most people in good relationships want to share what they have learned and have a hoard of funny stories of trials they overcame.

Until recently, young marriages would be actively supported by local religious leaders, priests, imams, and rabbis with years of experience. If you want help, but don't want to pay a shrink or therapist, try attending a house of worship and finding a local leader you respect.

Talk to that person about coping mechanisms for when times get hard. Most religious leaders are delighted to have new members join their congregation, and you don't have to be a fervent believer to get benefit out of belonging to a community where the members follow the golden rule.

Think about couples whose connection you find compelling. What keeps them engaged, and how much flexibility do they allow each other? A friend of mine once sat next to a couple on a plane ride. Watching them settle in for the journey together made her realize how miserable she was in her own marriage. And she's now in a far happier relationship.

Businesspeople talk feverishly about the legendary founders and entrepreneurs of big American companies who inspire them. Who inspires you in relationships?

When your aunt or uncle asks at Thanksgiving if you have found someone special yet, don't succumb to the full-body eye roll and change the subject. Ask instead how she and he knew they were right for each other, or how they decided to get married. If you know they went through a difficult time—perhaps they overcame illness or bankruptcy—ask them how they handled the pressure. Or if they're divorced, ask them when they knew it was time to call it quits.

My point is this: Most people—even those who annoy you—can offer wisdom on relationships based on their own experience. You don't have to take their advice. But other people's perspectives are useful. Especially those who have been together for many years and managed to survive without murdering each other.

Under your list of relationship role models, answer the following questions:

What is it about these couples that inspires you?

How do they treat each other?

What do you want to replicate in your relationship?

Now here is the fun part: interview them.
Tell them you are doing this as a personal assignment.

They will be flattered that you chose them, so you will be off to a good start. And then ask them all the questions you want to know, starting with "What is your secret?" Some may have their own equivalent of BIC. Others may say, "Don't sweat the small stuff!" Or "It has not always been this easy." Listen to their stories, ask more questions, and take lots of notes. We need mentors, and often they are at our fingertips.

Rule #15

Life is a feast. Take your place at the table.

"We must love one another, or die," urged the poet W. H. Auden in a line that has stuck with me since high school. And I like the poignancy of love lost in the Rascal Flatts song: "I should have stolen every moment / Now there's a page with not enough on it."

And let me add a final nod to Michael Pollan. Don't date someone your grandmother wouldn't want for you. Date someone who is kind and reliable. Who loves you for who you are. Who will treat you with respect. Who will love you, through thick and thin, in sickness and in health. Someone who makes you feel excited, happy, gorgeous, smart, and someone who you trust.

But don't go into a relationship already worried about "till death do us part." Thinking that this person will be the one and only is too much pressure. It's like starting as an intern and already worrying about making partner. Maybe that is why so many women have a hard time finding the One. We assume that when he comes along he will solve all our problems. It's the Cinderella complex. The Mr. Right terminology makes me wince. It suggests that there is only one solution.

But the truth is, there isn't. We ignore opportunities that could blossom into something worthwhile because we live in a Western culture that teaches us that there is one big love. I prefer

Helen Fisher's idea of "slow love." Like the slow food movement, where the best food is grown locally and eaten in season, slow love is an organic and healthy process that involves three pillars: lust, romance, and attachment. The first two can happen either first or second. In other words, you can fall in love with someone before you have sex, or you can have sex first and then fall in love. But the attachment stage, that third piece, is what you need for a long-term partner. That takes time, and as much as you would like to, you can't speed it up.

Along those lines, there is also no such thing as perfect when it comes to love—in you or your partner. Embrace imperfection and find a mate who loves you for yours. And know when you start dating that this person may not be the person you are with for the rest of your life. People change. *You* will change.

Embrace imperfection and find a mate who loves you for yours.

"People used to marry and have sex for the first time that night, and then stop having sex with other people," Esther Perel says. "They used to marry or choose someone for life; they didn't really have any exit option. Today, you can start the whole thing from scratch at sixty and even do it for the first time. That is a whole set of options that never existed before. Monogamy used to be one person for life, and today monogamy is one person at a time."

Today, 40 to 50 percent of US marriages end in divorce, according to the American Psychological Association. The Pew Research Center has found that marriage rates have declined in the US—one in ten adults ages twenty-five and older was not married in 1960; in 2012, it was one in five. Yet most people who are looking for a partner say that they want to find the one person they will spend the rest of their life with, according to Fisher's Match.com study. I wonder if that is simply unrealistic. What if you found someone who makes great sense right now? And for this next stage of your life?

"Our expectations have only risen," Perel says. "We still want everything we wanted from traditional relationships—companionship, economic support, family life, children, and legacy. But on top of it, we want our partner to be our best friend, trusted confidant, passionate lover, and intellectual equal, as well as the best parent." We live twice as long today as we did a hundred years ago, which puts that much more pressure on this one person. "The dominant romantic model of the day is that we are asking of one person what once an entire village used to provide," Perel adds. "And this belief that marriage will give you access to all these other things that are going to make you happy. Marriage isn't meant to be the source of happiness."

The happiness bit is up to you.

And you need to live the biggest life you can.

Sylvia Ann Hewlett conducted a study, called "Women Want Five Things," among successful working women and found that even more than marrying or having a child, women wanted to feel "exhilarated by their lives." Hewlett calls it "flourishing," and whether you want to find a partner to marry and have a child with or simply to find someone to share your life with, who doesn't want to flourish?

What makes you happy? If it's bird-watching or bowling, do it. Or if you don't know, try it. Yoga or tango. The local shooting range, a demolition derby, or Russian literature. Do all of it while you continue to search for love. And make that search a conscious one. It is too important to simply *hope* that something will happen.

When I was at *Cosmo*, alongside raises and orgasms, "How do you get him to commit?" was another frequently asked how-to question from readers. For those women who are dating the guy who won't say "I love you" or move in, move on already. Nothing is more depressing than having to persuade someone of your charms when they're not convinced. Any fear of having that conversation signifies that it is not an equal relationship.

If it is clearly not working, be brave and end it.

"We have this extension of what I call the pre-commitment, or commitment lite," Fisher says. "Marriage used to be the beginning of a relationship with someone, and now it is the finale. As a result you have plenty of time to try people out before you commit. Do it. If it does not work after three months, don't do it anymore. It's not hard to break up with people."

Look at it this way: If you were working on a deal, and the client was refusing to sign the papers or not returning your calls, you would know there was something wrong. You would not invent ridiculous reasons as to why. You would understand the silence and refusal means they have lost interest, gotten cold feet, and you would move on and look elsewhere for a new business lead. Likewise, do yourself a favor and be honest if you are waiting for the text or call and it has not yet arrived and it should have. Yes, it's annoying, disappointing, humiliating when someone doesn't return the ball you have thrown at them, but do not overinvest in someone who is not reciprocating.

Matthew Hussey, *Cosmo*'s relationship expert, has an excellent rule: "Only invest in someone who is equally investing in you." You want to create forward momentum so your love life does not stand or fall on him or her. You want equality here. This is the great benefit of dating apps. You can find someone else. So suck it up and move on if what seems like a promising lead dries up. Yes, I know, it takes courage to talk to the person in the distant cube or in front of you in the coffee line. If the thought of switching off your phone is too panic-inducing, then you can use it as an introducing tool, but in this real-life scenario: Turn to the guy behind you and reference the news flash that just ran across your screen. Or simply ask, "Do you know the Wi-Fi password?" Any excuse to actually look him in the eye and make a real connection, rather than disappearing into your phone and preventing him, and anyone else in that coffee shop or taxi line or whatever real-life line, from approaching you and starting a conversation.

These casual interactions are disappearing because of our focus on the phone. This means that we are missing the little entrées to meeting people. Plus, we get less practice at them and so find these first conversations—a necessary start to any relationship—intimidating when they don't need to be. So start practicing.

The added benefit of a genuine encounter is you can tell immediately if he is worth pursuing, whereas if you met him online, you would have no idea. It might take 250 texts to make a coffee date, and then you realize there's no way. But if you are standing opposite each other, filling up at the gas station, or you regularly pass each other while you're walking your dog, you can already suss out whether there might be a connection. Think of the time you will save just by hearing his voice.

And don't forget to take a few risks.

It's easy to box yourself in with a set of rules about what your future partner should or shouldn't be.

A friend of mine once freaked out because one morning her new boyfriend had left her a Post-it note that said, "We must certainly meet up again." The problem was that he had spelled the world *certainly* with an *S*. Yes, as in *sertainly*.

"I went to Princeton. I can't go out with someone who can't spell!" she shouted.

But she liked him and stuck it out and realized that they had so much in common—including a love of food and cooking—and now twenty years later, they have three children together.

Put your preconceptions aside. It could lead to an idea of happiness that you did not know existed.

Prepare to surprise yourself.

I have only two real requirements for the people I work with. Turn up on time, and have fun. You could apply the same to first dates.

We've established that dating can be hard—but it does not have to be. Attitude is everything. I love Hewlett's description of flourishing. It's also about taking the pressure off the hunt for this one person who will solve all your problems. The princess waiting for her prince. Get up on your own horse. And as you learn to ride, look for someone who can trot, canter, and eventually gallop alongside you.

The wedding industry adds to the pressure of finding the "one and only." Marriage has become this mirage in our culture where more focus is placed on the party—which can cost tens and sometimes hundreds of thousands of dollars—than on the union being celebrated. Women spend more time with their wedding planners figuring out what type of canapés to serve, wine to pour, and flowers to hold, than with a couples counselor or religious leader to prepare for what happens after the honeymoon is over, the presents are opened, and the thank-you notes are sent.

And then you are left with the real work of a marriage. A partnership.

We don't talk enough about the benefits of a good relationship. Study after study proves that solid, loving relationships that make you feel wanted, respected, relaxed, and safe are essential to both happiness and good health. Harvard University began a study of 268 men, all age nineteen, in the late 1930s and has been tracking them, and their families, ever since. During the years since, that control group has grown. Robert Waldinger, the current executive director of the study, reports that those men in "low-conflict relationships" had happy marriages and actually lived longer. High-conflict marriages turn out to be very bad for our health, "perhaps worse than getting divorced," Waldinger shared in his TED Talk, which has been viewed more than seven million times. Meanwhile, "living in the midst of good, warm relationships is protective."

And remember, it's not retro to want to find a partner and have kids. During the immediate, horrific aftermath of 9/11, one piece stood out to me in the cacophony. Writing in the *Guardian*, the wonderful British novelist Ian McEwan detailed the story of a husband who missed the panicked call from his sobbing wife and was forced to replay for news crews her final message left on the family answering machine. The fire was raging, and there was no way out of the tower.

She was calling to say goodbye.

"There was really only one thing for her to say," he writes. "Those three words that all the terrible art, the worst pop songs and movies, the most seductive lies, can somehow never cheapen. I love you. She said it over and again before the line went dead."

"I love you."

I love you.

I love you.

I love you.

I love you.

Love.

The food of life. We crave it, we depend on it, and in so many ways, it is all that matters in the end. Whom you loved and who loved you back. Which is why we must choose it wisely and well.

My final rule is simple. Life is a feast. So take your place at the table and love someone who knows you're special and loves you back. You deserve it.

Acknowledgments

First and foremost, huge thanks to Liz Welch, my coauthor, who made every moment of writing and researching this book with her a pleasure. Her diligence, curiosity, extreme smarts, organization, and attention to deadlines were a gift. I can't wait to do this with her again.

Huge thanks also to Jonathan Burnham at HarperCollins, for trusting me to write this and pairing me with the wonderful editor Jennifer Barth, whose extreme intelligence and unobtrusive manner nudged me at crucial times in a much better direction.

To the cover designer Robin Bilardello and the whole HarperCollins team: Tina Andreadis, Leslie Cohen, Stephanie Cooper, Doug Jones, and Erin Wicks.

To Bob Barnett who educated me on the New York publishing business amid many chocolate malts from Shake Shack.

For sharing their remarkable insight, wisdom, and research so generously: David Adams; Mary Aiken, PhD; Gail Dines, PhD; Leah Fessler; Helen Fisher, PhD; Cindy Gallop; Justin Garcia, PhD; Elizabeth Gregory, PhD; Sarah Hepola; Sylvia Ann Hewlett; David Jernigan, PhD; Steve Kardian; Marcelle Karp; Ian Kerner, PhD; Logan Levkoff; Wednesday Martin, PhD; Esther Perel; Sean Rad; Tanya Selvaratnam; Rachel Sklar; Julie Spira; Jean Twenge, PhD; Sharon Wilsnack, PhD; and Whitney Wolfe.

And to the fantastic editors at *Cosmopolitan* and *Marie Claire*, with whom I have argued, learned, raged, and fallen about laughing. They are all feisty, funny, and fabulous: Joyce Chang, Leslie Yazel, Sara Austin, Laura Brounstein, Marina Khidekel, Abigail Pesta, Lea Goldman, Emily Johnson, Angela Ledgerwood, Lucy Kaylin, Anne Fulenwider, Julie Vadnal, Emily Johnson, Holly Whidden, Michele Promaulayko, Leah Wyar, Aya Kanai, Tiffany Reid, Amy Odell, Lori Fradkin, Kate Lewis, Matthew Hussey, Steven Brown, James Worthington DeMolet, Logan Hill, and Sergio Kletnoy.

To my brilliantly patient and resourceful assistant, Heather Passaro.

And to two of the best publishers in the business, Nancy Berger and Donna Lagani.

To our fact-checker Jennifer Kelly Geddes, transcriber Martha Sorren, and all the early readers, including Gabriella Cirelli and Olivia Winn.

To my corporate Hearst colleagues David Carey and Michael Clinton, whose daily support, smarts, and sense of humor makes work such a pleasure.

To Dave Bernad, the executive producer, and the entire team at Freeform, led by Karey Burke, Tom Ascheim, and Simran Sethi, who encouraged everyone at *The Bold Type* to be as real and unapologetic as possible around the issues impacting young women.

And for the genius casting of Melora Hardin, Katie Stevens, Meghann Fahy, Aisha Dee, Sam Page, and Matt Ward, whose talent and craft have brought many of these complicated issues to the screen so effectively in the show.

And then to my friends and fellow authors who encouraged me and energetically lent their suggestions as I moaned to them about spending my weekends writing: Jane Thynne, Daisy Goodwin, Georgina Godwin, Colleen DeCourcy, Sheryl Sandberg, Arianna Huffington, Mika Brzezinski, Nell Scovell, Deb Spar, and Susan Mercandetti for pushing me to get on with it.

To Mum, Dad, Liz, Con, Mary, Peter, Thomas, and Hugo.

And a special thanks to the many, many people who have trusted me over the years with their personal stories and frustrations and longings, and who for privacy reasons didn't want their real names published. As I said in the foreword, this book is for you.

Notes

FOREWORD

ix one book stands out: Michael Pollan, *Food Rules: An Eater's Manual* (New York: Penguin, 2011).

RULE #1: ESTABLISH YOUR IDEAL LOVE WEIGHT.

4 "looking-glass self" was first coined: Charles Horton Cooley, "The Looking-Glass Self," in *Social Theory: The Multicultural, Global, and Classic Readings*, ed. C. Lemert (Philadelphia, PA: Westview Press), 189.

4 "investing in trying to understand our 'self'": Mary Aiken, *The Cyber Effect: A Pioneering Cyberpsychologist Explains How Human Behavior Changes Online* (New York: Speigel and Grau, 2016).

RULE #2: CLEAR OUT YOUR CUPBOARDS AND SWEEP THE FRIDGE.

12 "It's so hard for women to admit that they want this": Helen Fisher, *Anatomy of Love: A Natural History of Mating, Marriage, and Why We Stray* (New York: W. W. Norton, 2016).

15 College undergraduate intake: US Department of Education, National Center for Education Statistics, *The Condition of Education 2017* (NCES 2017–144), Table 303.10, https://nces.ed.gov/programs/digest/d13/tables /dt13_303.10.asp.

RULE #3: BEGIN A DATING DETOX TO RESET YOUR METABOLISM.

25 Leah Fessler, who based her senior thesis: Leah Fessler, "A Lot of Women Don't Enjoy Hookup Culture—So Why Do We Force Ourselves to Participate?" *Quartz*, May 17, 2016, https://qz.com/685852 /hookup-culture/.

RULE #4: THE TREADMILL WON'T RUN ON ITS OWN. CLIMB ON AND PRESS START.

35 Esther Perel—the brilliant couples therapist: Esther Perel, *The State of Affairs: Rethinking Infidelity* (New York: HarperCollins, 2017).

37 number of people ages eighteen to twenty-four who use them: Aaron Smith and Monica Anderson, "5 Facts About Online Dating," Pew Research Center, February 2017, http://www.pewresearch.org/fact-tank/2016/02/29/5-facts-about-online-dating/.

37 has nearly tripled from 10 percent: Ibid.

38 "In cyberpsychology, we refer to Walther's Theory": Joseph P. Walther, "Computer-Mediated Communication: Impersonal, Interpersonal, and Hyperpersonal Interaction," *Communication Research* 23, no. 1 (1996).

39 what Aiken calls the "stranger on the train syndrome": Sabina Misoch, "Stranger on the Internet: Online Self-Disclosure and the Role of Visual Anonymity," *Computers in Human Behavior* 48 (2015): 535–41.

39 Online, Aiken says, that level of self-disclosure doubles: Chris Fullwood, Mike Thelwall, and Sam O'Neill, "Clandestine Chatters: Self-Disclosure in U.K. Chat Room profiles," *First Monday* 6, no. 5 (May 2011), https://firstmonday.org/ojs/index.php/fm/article/view/3231/2954.

39 the sixfold increase in sexual assault: National Crime Agency, *Emerging New Threat in Online Dating: Initial Trends in Internet Dating Initiated Serious Sexual Assaults*, February 2017, http://www.nationalcrimeagency.gov.uk/publications/670-emerging-new-threat-in-online-dating-initial-trends-in-internet-dating-initiated-serious-sexual-assaults/file.

40 71 percent of those reported assaults: Ibid.

40 one half of all sexual assaults involve alcohol: Antonia Abbey, Tina Zawacki, Philip O. Buck, A. Monique Clinton, and Pam McAuslan, "Alcohol and Sexual Assault," National Institute on Alcohol Abuse and Alcoholism, https://pubs.niaaa.nih.gov/publications/arh25-1/43-51.htm.

43 Tinder and Match are among the top ten: Ben Gray, "The Best Dating Apps to Get Before Valentine's Day," *ARC Report*, February 2016, https://arc.applause.com/2016/02/10/best-and-worst-dating-apps-for-2016/.

44 Clearly women agree: Whitney Wolfe, "Bumble Celebrates First Movers for International Women's Day," The Beehive (blog), Bumble.com, March 7, 2017, http://blog.bumble.com/bumbleblog/international-womens-day2017.

RULE #5: CHOOSE THE RIGHT RECIPES FOR YOUR DATING TYPE.

51 15 percent of American adults have dated online: Aaron Smith, "15% of American Adults Have Used Online Dating Sites or Mobile

Dating Apps," Pew Research Center, February 11, 2016, http://www
.pewinternet.org/2016/02/11/15-percent-of-american-adults-have-used
-online-dating-sites-or-mobile-dating-apps/.

69 *New Super Power for Women*: Steve Kardian, *The New Super Power
for Women: Trust Your Intuition, Predict Dangerous Situations, and Defend
Yourself from the Unthinkable* (New York: Touchstone, 2017).

72 One in twenty-five people: Amanda Lenhart, Michelle Ybarra,
and Myeshia Price-Feeney, "Nonconsensual Image Sharing: One in
25 Americans Has Been a Victim of 'Revenge Porn,'" Data & Society Research Institute, December 13, 2016, https://datasociety.net
/pubs/oh/Nonconsensual_Image_Sharing_2016.pdf.

73 16 percent told McAfee: "Love, Relationships and Technology," McAfee,
February 4, 2014, https://promos.mcafee.com/offer.aspx?id=605366.

73 "changes the risk profiles": Justin Garcia, Amanda N. Gesselman,
Shadia A. Siliman, Brea L. Perry, Kathryn Coe, and Helen E. Fisher,
"Sexting Among Singles in the USA: Prevalence of Sending, Receiving,
and Sharing Sexual Messages and Images," *Sexual Health* 13, no. 5 (July
2016): 428–35. http://archive.news.indiana.edu/releases/iu/2016/08/sex
ting-research.shtml.

RULE #6: YOU WON'T GET SKINNY BY EATING THE SAME OLD SH*T.

85 42 percent of people ages twenty-five to twenty-nine: D'Vera Cohn,
Jeffrey S. Passel, Wendy Wang, and Gretchen Livingston, "Barely Half
of US Adults Are Married—A Record Low: New Marriages Down 5%
from 2009 to 2010," Pew Research Center, December 14, 2011, http://
www.pewsocialtrends.org/2011/12/14/barely-half-of-u-s-adults-are
-married-a-record-low/.

86 examines the rise in the number of financially independent and
single women: Rebecca Traister, *All the Single Ladies: Unmarried Women
and the Rise of an Independent Nation* (New York: Simon & Schuster,
2016).

88 the annual "Singles in America" study: Helen Fisher and Justin R.
Garcia, "Singles in America: 2017," Match.com, http://www.singlesin
america.com/2017/.

**RULE #7: STOP WITH THE COMFORT FOODS. IT'S OKAY TO BE A LITTLE
HUNGRY.**

98 it unleashes all the feels: A. Argiolas and M. R. Melis, "Dopamine
and Sexual Behavior," *Neuroscience & Biobehavioral Reviews* 19, no. 1
(Spring 1995): 19–38, https://www.ncbi.nlm.nih.gov/pubmed/7770195.

101 "Dunbar's number dictates": R. I. M. Dunbar, "Coevolution of Neo-cortical Size, Group Size and Language in Humans," *Behavioral and Brain Sciences* 16 (1993): 681–735.

RULE #8: ALCOHOL IS NOT A FOOD GROUP. RESPECT YOUR LIMITS.

109 Answer: Alcohol: "Drugs/Narcotics and Alcohol Involvement," Federal Bureau of Investigation, National Incident-Based Reporting System, Uniform Crime Reporting, https://ucr.fbi.gov/nibrs/2012/table-pdfs/drugs-narcotics-and-alcohol-involvement-by-offense-category-2012.

109 one half of all reported sexual assaults: Antonia Abbey, Tina Za-wacki, Philip O. Buck, A. Monique Clinton, and Pam McAuslan, "Al-cohol and Sexual Assault," National Institute on Alcohol Abuse and Alcoholism.

110 a more active stomach enzyme: Harold Franzen, "Enzyme Lack Lowers Women's Alcohol Tolerance," *Scientific American*, April 16, 2001.

110 defines binge drinking for women: National Institute of Health, Al-cohol Facts and Statistics on Alcohol Abuse and Alcoholism, https://www.niaaa.nih.gov/alcohol-health/overview-alcohol-consumption/moderate-binge-drinking.

111 in over her head using alcohol to help find love: Sarah Hepola, *Blackout: Remembering the Things I Drank to Forget* (New York: Grand Central Publishing, 2015).

111 NIAAA reported that of the 60 percent of all college students who drank alcohol: National Institute of Health, Alcohol Facts and Statistics on Alcohol Abuse and Alcoholism, http://www.healthdata.org/news-release/heavy-drinking-and-binge-drinking-rise-sharply-us-counties, https://www.collegedrinkingprevention.gov/media/college drinkingFactSheet.pdf.

111 only risen 4.9 percent among men...[and] 17.5 percent among women: Laura Dwyer-Lindgren, Abraham D. Flaxman, Marie Ng, Gillian M. Hansen, Christopher J. L. Murray, and Ali H. Mokdad, "Drinking Pat-terns in US Counties from 2002 to 2012," *American Journal of Public Health* 105, no. 6 (June 2015): 1120–1127, May 13, 2015, http://ajph.aphapub lications.org/doi/abs/10.2105/AJPH.2014.302313.

112 white women ages thirty-five to fifty-four: Kimberly Kindy and Dan Keating, "For Women, Heavy Drinking Has Been Normalized. That's Dangerous," *Washington Post*, December 23, 2016, https://www.washington post.com/national/for-women-heavy-drinking-has-been-normalized-thats-dangerous/2016/12/23/0e701120-c381-11e6-9578-0054287507db_story .html?utm_term=.08b3f8809528.

113 women felt more attractive while buzzed: Megan E. Patrick and Jennifer L. Maggs, "Does Drinking Lead to Sex? Daily Alcohol-Sex Behaviors and Expectancies Among College Students," *Psychology of Addictive Behaviors* 23, no. 3 (September 2009): 472–81.

113 82 percent of college students who had unwanted sex: Aaron White and Ralph Hingson, "The Burden of Alcohol Use: Excessive Alcohol Consumption and Related Consequences Among College Students," *Alcohol Research: Current Reviews* 35, no. 2, https://pubs.niaaa.nih.gov /publications/arcr352/201-218.htm.

114 47 percent of college-aged women who were raped: "Alcohol Facilitated Sexual Assault," Virginia State University, 1995, http://www.vsu .edu/counseling/substance-abuse/alcohol-and-sexual-assault.php.

116 the rules around advertising alcohol: "Code of Responsible Practices for Beverage Advertising and Marketing," Distilled Spirits Council, http://www.discus.org/assets/1/7/May_26_2011_DISCUS_Code_Word _Version1.pdf.

116 "Eighty-five percent of all drinkers": "Underage Drinking," National Institute of Health and Human Services, October 2010, https:// report.nih.gov/NIHfactsheets/ViewFactSheet.aspx?csid=21.

117 Between 2005 and 2012: "Binge drinking," Alcohol and Public Health, Centers for Disease Control and Prevention, 2017, https://www .cdc.gov/alcohol/fact-sheets/binge-drinking.htm.

117 "substance abuse in girls than boys": Ramesh Shivani, R. Jeffrey Goldsmith, and Robert M. Anthenelli, "Alcoholism and Psychiatric Disorders," National Institute on Alcohol Abuse and Alcoholism, https://pubs.niaaa.nih.gov/publications/arh26-2/90-98.htm.

120 sentenced to nine years in prison: John Futty, "Woman Who Livestreamed Girl's Rape Sentenced to Nine Months," *Columbus Dispatch*, February 13, 2017, http://www.dispatch.com/news/20170213/woman -who-live-streamed-girls-rape-sentenced-to-nine-months.

120 her friend: "Statistics About Sexual Violence," National Sexual Violence Resource Center, http://www.nsvrc.org/sites/default/files/pub lications_nsvrc_factsheet_media-packet_statistics-about-sexual-vio lence_0pdf.

RULE #9: HOOKUPS ARE LIKE FRENCH FRIES.

124 millennials are more than twice as likely: Jean Twenge, Ryne A. Sherman, and Brooke E. Wells, "Sexual Inactivity During Young Adulthood Is More Common Among US Millennials and iGen: Age, Period, and Cohort Effects on Having No Sexual Partners After Age 18," *Archives of Sexual Behavior* 46, no. 2 (February 2017): 433–40.

125 "a host of powerful feelings": Navneet Magon and Sanjay Kalra, "The Orgasmic History of Oxytocin: Love, Lust, and Labor," *Indian Journal of Endocrinology and Metabolism* 15, suppl. 3 (September 2011): S156–S161, doi: 10.4103/2230-8210.84851.

126 "they really want a relationship as well": Justin R. Garcia, Chris Reiber, Sean Massey, and Ann Merriwether, "Sexual Hookup Culture: A Review," *Review of General Psychology* 16, no. 2 (2012): 161–76, https://www.apa.org/monitor/2013/02/sexual-hookup-culture.pdf.

126 "You're invoking all senses": Ian Kerner, *She Comes First: The Thinking Man's Guide to Pleasuring a Woman* (New York: HarperCollins, 2010).

127 That number grew to 16 percent: Elizabeth Armstrong, Paula England, and Alison C. K. Fogarty, "Accounting for Women's Orgasm and Sexual Enjoyment in College Hookups and Relationships," *American Sociological Review* 77, no. 3 (June 1, 2012): 435–62, https://doi.org/10.1177/0003122412445802.

133 forthcoming book on female infidelity: Wednesday Martin, PhD, *Untrue: Why Nearly Everything We Believe About Women, Lust, and Infidelity Is Wrong and How the New Science Can Set Us Free* (New York: Little Brown, 2018).

134 A 2014 study: Samantha J. Dawson and Meredith L. Chivers, "Gender Differences and Similarities in Sexual Desire," *Current Sexual Health Reports* 6, no. 4 (December 2014): 211–19, https://link.springer.com/article/10.1007/s11930-014-0027-5.

RULE #10: PORN IS LIKE CHEWING GUM—ALL ARTIFICIAL FLAVOR.

142 "Porn is to sex": Gail Dines, *Pornland: How Porn Has Hijacked Our Sexuality* (Boston: Beacon Press, 2010).

144 2.6 million visitors per *hour*: "Pornhub's 2016 Year in Review," Pornhub Insights, January 4, 2017, https://www.pornhub.com/insights/2016-year-in-review.

144 referred to as the Kinsey Report of our time: Maureen O'Connor, "Pornhub is the Kinsey Report of our Time," *New York*, June 11, 2017, https://www.thecut.com/2017/06/pornhub-and-the-american-sexual-imagination.html.

144 age data starts at eighteen: "Pornhub's 2016 Year in Review."

144 regularly visit porn sites: Ibid.

145 porn consumption lowered commitment: Nathaniel M. Lamert, Sesen Negash, Tyler F. Stillman, Spencer B. Olmstead, and Frank D. Fincham, "A Love That Doesn't Last: Pornography Consumption and Weakened Commitment to One's Romantic Partner," *Journal of Social and Clinical Psychology* 31, no. 4 (2012): 410–38.

145 porn-free relationships are stronger: Peg Streep, "What Porn Does to Intimacy: 3 Studies Find That Explicit Material Can Do More Harm Than Most People Think," *Psychology Today*, July 16, 2014, https://www.psychology today.com/blog/tech-support/201407/what-porn-does-intimacy?page=1.

146 started her website MakeLoveNotPorn: www.makelovenotporn.com.

147 Other terms trending: "Pornhub's 2016 Year in Review."

RULE #11: STICK TO NATURAL SUGARS. SUBSTITUTES ARE BAD FOR YOUR HEALTH.

157 One in three women in the US has been physically abused: "National Statistics," National Coalition Against Domestic Violence, http://ncadv.org/learn-more/statistics.

157 half the time by a man she knows: Ibid.

158 "lacks empathy": narcissistic personality disorder as defined in the *Diagnostic and Statistical Manual of Mental Disorders*, "DSM-IV and DSM-5 Criteria for the Personality Disorders," American Psychiatric Association, 2012, http://www.psi.uba.ar/academica/carrerasdegrado/psicologia /sitios_catedras/practicas_profesionales/820_clinica_tr_personalidad _psicosis/material/dsm.pdf.

158 narcissistic personality disorder affects men more greatly: Emily Grijalva, Daniel Newman, et al., "Gender Differences in Narcissism—A Meta-Analytic Review," *Psychological Bulletin* 141, no. 2 (March 2015): 261–310, doi: 10.1037/a0038231.

162 argues that social media exacerbates: Jean M. Twenge, *Generation Me: Why Today's Young Americans Are More Confident, Assertive, Entitled—and More Miserable Than Ever Before* (New York: Free Press, 2006).

RULE #12: TRUST YOUR GUT AND PROTECT YOURSELF WITH PROBIOTICS.

174 "positive emotions, positive social connections": Bethany Kok, Kimberly Coffey, Michael Cohn, et al., "How Positive Emotions Build Physical Health," *Psychological Science* 24, no. 7 (July 1, 2013): 1123–32, https://doi.org/10.1177/0956797612470827.

175 witnessing acts of kindness produces oxytocin: Jill Suttie, "How Our Bodies React to Seeing Goodness," *Greater Good Magazine*, University of California, Berkeley, May 12, 2015, https://greatergood.berkeley .edu/article/item/how_our_bodies_react_human_goodness.

175 known as the "helper's high": "Giving of Your Gifts," Emory Faculty Staff Assistance Program, http://www.fsap.emory.edu/workplace-resources/well nessmatters/giving-your-gifts.html.

RULE #13: SET YOUR OWN "BEST BEFORE" DATE.

185 peak fertility is around age twenty-six: Sarah Graham, "Study Shows Fertility Decline Begins in Late 20s," *Scientific American*, May 1, 2002, https://www.scientificamerican.com/article/study-shows-fertility-dec/.

186 Cameron Diaz announced: Tom Chiarella, "Cameron Diaz Is the Best She's Ever Been," *Esquire*, August 2014, http://www.esquire.com/entertainment/interviews/a32839/cameron-diaz-story-0814/.

186 reasons she decided not to have children: Sezín Koehler, "8 Reasons Why I Am Not Having Children," *HuffPost*, September 15, 2014, http://www.huffingtonpost.com/sezin-koehler/8-reasons-why-im-not-having-children-childfree_b_5705311.html.

186 Polly Vernon, an editor at the *Guardian*: Polly Vernon, "Why I Don't Want Children," *Guardian*, February 7, 2009, https://www.theguardian.com/lifeandstyle/2009/feb/08/motherhood-children-babies1.

187 Tanya Selvaratnam's book: Tanya Selvaratnam, *The Big Lie: Motherhood, Feminism, and the Reality of the Biological Clock* (Amherst, NY: Prometheus Books, 2014).

188 40 percent of women earning over $50,000: Sylvia Ann Hewlett, *Creating A Life: Professional Women and the Quest for Children* (New York: Miramax, 2002).

189 By forty, that decreases: American Congress of Obstetricians and Gynecologists, Committee Opinion, no. 589, March 2014, https://www.acog.org/Resources-And-Publications/Committee-Opinions/Committee-on-Gynecologic-Practice/Female-Age-Related-Fertility-Decline.

189 in vitro gametogenesis, the latest development in the infertility field: Tamar Lewin, "Babies from Skin Cells? Prospect Is Unsettling to Some Experts," *New York Times*, May 16, 2017, https://www.nytimes.com/2017/05/16/health/ivg-reproductive-technology.html?_r=0.

190 Just under 40 percent of IVF cycles: "Success Rates," Society for Assisted Reproductive Technology, http://www.sart.org/SART_Success_Rates/.

190 Only 40 percent of all IVF rounds: Ellie Kincaid, "The Success Rates from IVF Are Nowhere Near What People Think," *Business Insider*, Australia, May 29, 2015, https://www.businessinsider.com.au/in-vitro-fertilization-ivf-success-rates-2015-5.

191 70 percent of all women who sought some form of assisted reproductive technology: "FAQ: What is the cost of IVF?" Society for Assisted Reproductive Technology, http://www.sart.org/patients/frequently-asked-questions/.

192 Alternative Parenting has pioneered: Alternative Parenting Center, http://www.alp.org.il/page/142.html.

193 interviewed more than one hundred women: Elizabeth Gregory, *Ready: Why Women Are Embracing the New Later Motherhood* (New York: Basic Books, 2007).

196 including a basal antral follicle count: See for example http://www .rmact.com/getting-started/fertility-testing/basal-antral-follicle-count.

196 hysterosalpingogram, which is an X-ray: "Hysterosalpingogram," WebMD, http://www.webmd.com/women/hysterosalpingogram-21590#1.

197 including hyperstimulation syndrome: "Ovarian Hyperstimulation Syndrome," Mayo Clinic, http://www.mayoclinic.org/diseases-conditions /ovarian-hyperstimulation-syndrome-ohss/home/ovc-20263580.

RULE #15: LIFE IS A FEAST. TAKE YOUR PLACE AT THE TABLE.

216 Helen Fisher's idea of "slow love": Helen Fisher, "Casual Sex May Be Improving America's Marriages: One-Night Stands and Friends with Benefits Are Just What Your Brain Ordered," *Nautilus*, March 5, 2015, http://nautil.us/issue/22/slow/casual-sex-is-improving-americas -marriages.

216 40 to 50 percent of US marriages end in divorce: "Marriage and Divorce," American Psychological Association website, http://www.apa .org/topics/divorce/.

216 marriage rates have declined in the US: Pew Research Center, http:// www.pewsocialtrends.org/2014/09/24/record-share-of-americans -have-never-married/.

217 one person they will spend the rest of their life with: "Singles in America," Match.com, https://www.multivu.com/players/English/8024551 -match-7th-annual-singles-in-america-study/.

217 Hewlett calls it "flourishing": Sylvia Ann Hewlett, "Women Want Five Things," Center for Talent Innovation, December 9, 2014, http:// www.talentinnovation.org/publication.cfm?publication=1451.

221 that control group has grown: Harvard Second Generation Study, http://www.adultdevelopmentstudy.org.

221 those men in "low-conflict relationships": Liz Mineo, "Good Genes Are Nice, But Joy Is Better," *Harvard Gazette*, April 11, 2017, http:// news.harvard.edu/gazette/story/2017/04/over-nearly-80-years-harvard -study-has-been-showing-how-to-live-a-healthy-and-happy-life/.

221 "living in the midst of good, warm relationships": Ibid.

221 "There was really only one thing": Ian McEwan, "Only Love and Then Oblivion. Love Was All They Had to Set Against Their Murderers," *Guardian*, September 15, 2001, https://www.theguardian.com /world/2001/sep/15/september11.politicsphilosophyandsociety2.

Recommended Reading

In addition to all the books written by experts quoted throughout this book, here are more recommended titles:

Adichie, Chimamanda Ngozi. *We Should All Be Feminists*. New York: Anchor Books, 2015.

Ansari, Aziz. *Modern Romance: An Investigation*. New York: Penguin Press, 2015.

Duckworth, Angela. *Grit: The Power of Passion and Perseverance*. New York: Scribner, 2017.

Dunham, Lena. *Not That Kind of Girl: A Young Woman Tells You What She's Learned*. New York: Random House, 2014.

Grigoriadis, Vanessa. *Blurred Lines: Rethinking Sex, Power, and Consent on Campus*. New York: Houghton Mifflin Harcourt, 2017.

Levy, Ariel. *The Rules Do Not Apply: A Memoir*. New York: Random House, 2017.

Moran, Caitlin. *How to Be a Woman*. New York: Harper Perennial, 2012.

———. *Moranifesto*. New York: Harper Perennial, 2016.

Orenstein, Peggy. *Girls & Sex: Navigating the Complicated New Landscape*. New York: HarperCollins, 2016.

Paul, Pamela. *Pornified: How Pornography Is Damaging Our Lives, Our Relationships, and Our Families*. New York: Henry Holt, 2005.

Senior, Jennifer. *All Joy and No Fun: The Paradox of Modern Parenthood*. New York: HarperCollins, 2014.

Traister, Rebecca. *All the Single Ladies: Unmarried Women and the Rise of an Independent Nation*. New York: Simon and Schuster, 2016.

Turkle, Sherry. *Reclaiming Conversation: The Power of Talk in a Digital Age*. New York: Penguin Press, 2015.

———. *Alone Together: Why We Expect More from Technology and Less from Each Other*. New York: Penguin Press, 2011.

Valenti, Jessica. *Sex Object: A Memoir*. New York: HarperCollins, 2016.

Witt, Emily. *Future Sex: A New Kind of Free Love*. New York: Farrar, Straus and Giroux, 2016.

About the Author

JOANNA COLES is the chief content officer of Hearst Magazines and serves on the board of Snap Inc. She is the executive producer of *The Bold Type* on ABC's Freeform, a scripted show inspired by her life as a magazine editor, and starred in *So Cosmo* on E! Born in the United Kingdom, Coles was the New York correspondent for the *Guardian* and the *Times* of London before joining Hearst as editor in chief of *Marie Claire*. She was editor in chief of *Cosmopolitan* from 2012 to 2016. She lives in New York.